GLIMPSES into the OBVIOUS

for Betty and Fred

GLIMPSES into the OBVIOUS

Enjoy! from Florence

Sherman A. Wengerd

"Between early glimpses of reality and the fate that is inevitable lies the vast panopoly of unforgettable incidents that makes a life well-lived worthwhile. During a lifetime of observation as a child, adolescent and adult, a writer records experiences and thoughts which, through analysis, create strong impressions; and from such musings emerges a workable philosophy."

Sherman A. Wengerd

Goosefoot Acres Press
CLEVELAND, OHIO

Copyright © 1998, Florence Mather Wengerd.

All rights reserved. No part of this publication may be reproduced or transmitted in any form or by any means, electronic or mechanical, including photocopy, recording or any information storage and retreival system, except by a reviewer who may quote brief passages, without written permission.

ISBN 1-879863-54-5

Produced for the Wengerd family by, and additional copies available from:

Goosefoot Acres Press
Division of Goosefoot Acres, Inc.
P.O. Box 18016
Cleveland, OH 44118-0016
(216) 932-2145

Peter A. Gail, Ph.D. Publisher
Jane Takac, Editor

Library of Congress Cataloging-in-Publication Data

Wengerd, Sherman A. (Sherman Alexander), 1915-1995
Glimpses into the obvious/ by Sherman A. Wengerd [Jane Takac, editor]
p. cm.
ISBN 1-879863-54-5 (trade pbk.; alk. paper)
1. Wengerd, Sherman A. (Sherman Alexander, 1915-1995.
2. Petroleum geologists--United States--Biography. I. Takac, Jane.
II. Title.
TN869.2.W46A3 1998
550'.92--dc21
[B] 98-19857
CIP

For Florence

and our four children

Anne
Timothy
Diana
Stephanie

and their families

who gave me unexpected views of immortality

Sherman Alexander Wengerd
February 17, 1915 - January 28, 1995

PREFACE

Autobiographical tidbits are offered as preludes to one man's idealistic approach to life. The only practical way to find out why one thinks as one does is to look as rigorously as possible into one's own early background as well as the surroundings of our forebears.

This foray into the meaning of parts of a long life begins with stories from my childhood and youth and then continues with a search through the history of my background. Stories passed from generation to generation may lead to exaggeration of the good and denigration of the bad. Also, the few published details about early ancestors reveal times, conditions and milieux far different from those today in Switzerland and the America of the 18th and 19th centuries, as well as the early days of the 20th century.

To determine which facts are most important—whether humorous or tragic, dull or exciting—becomes a matter of choice before we "shuffle off this mortal coil," whether with a whimper or in a bright flash before oblivion.

Between early glimpses of reality and the fate that is inevitable lies the vast panoply of unforgettable incidents that makes a life well-lived worthwhile. During a lifetime of observation as a child, adolescent and adult, a writer records experiences and thoughts which, through analysis, create strong impressions; and from such musings emerges a workable philosophy.

In every sense, many of us who profess to be writers have as our mythical collaborators such practical wordsmiths as Sherwood Anderson, Ernest Hemingway, Robert Roark,

Paul Tabori, C. Northcote Parkinson, Richard Armour, James Michener, and Paul Theroux among others. Those who have gone before me provided shoulders on which I have stood in attempts to see more distant horizons. I am led to observe that humorous and serious quotations, even though inadvertently and slightly altered, tell us much about how to approach the good life.

So, this is a book about a great number of very interesting humans, most of whom have viewed life and their fellow humans with open-eyed amazement in constantly changing surroundings. Who has not found life filled with humor and pathos, misery and enthusiasm? But what of the future? Will it be better? More wars, depressions, enmities, riots, and terrorism? Of course! Humans are a perverse lot, bundles of contradictions. If we work toward the amity that will drive enmity off the face of our planet, and if we will but learn from history, the future will be much brighter than the past has been. So long as we thinking humans plan our lives as though we would live forever, and live each day as though it were our last, the world must become a better place.

Sherman A. Wengerd
Albuquerque, New Mexico, U.S.A.
October, 1994

ACKNOWLEDGMENTS

Many people have supplied their unseen hands and minds in the initiation and inspiration of these glimpses of life since the inception of research which began in Europe in July of 1960.

I am indebted to Cresap P. Watson, known as "Doc" by his hosts of friends. He gave me the title in 1960 with the admonition that I write the book that he knew he would not have time to write. An engaging conversationalist, Doc Watson was a good friend of Sydney Powers, a founder and early president of the American Association of Petroleum Geologists. In the early history of that association, Powers invited Doc to join AAPG as a full member despite the fact that Doc was a financier rather than a geologist. Doc Watson said of many a friend, "There is more gallantry than guile in that man!"

To Diana Wengerd Roach, my deep thanks for her efficient word processing production of this book (hard copy and diskette) as well as the intense editing she exerted on the original manuscript. The eagle eye of my wife, Florence Mather Wengerd, while typing several early sections of my tedious handwriting, is much appreciated. And thanks to my grandchildren for providing several illustrations for the book.

All errors are my responsibility—yet it is at times difficult for editors and critics to remember that some of the most engaging modern writing involves incomplete sentences, dangling participles, split infinitives, and shocking words which are the everyday menu of conversation.

CONTENTS

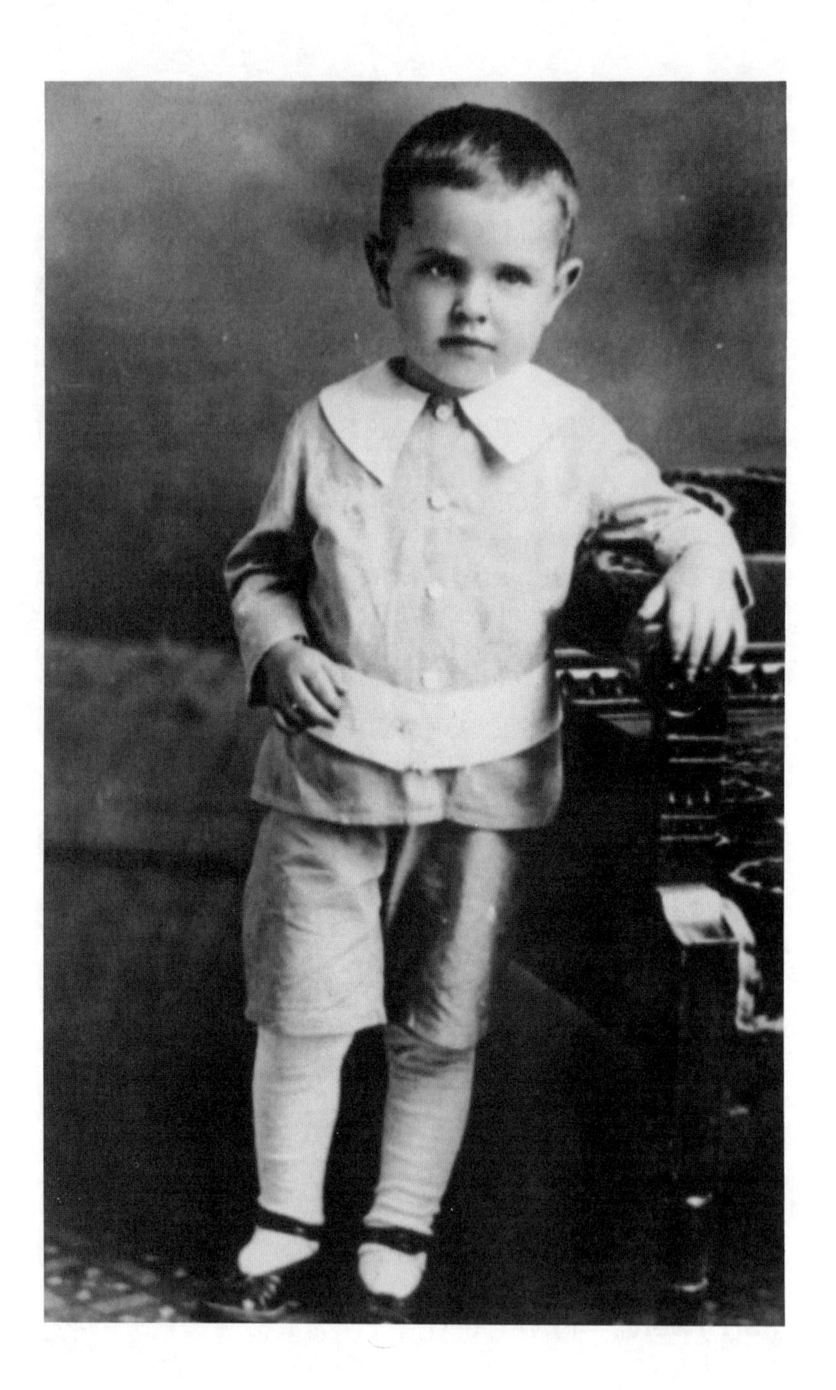

Part One

LEGACIES OF AN EARLIER DAY

A Boy in Ohio
Never a Dull Moment
Nicknames
A Field Next to School
The Standing Tree
The Chestnut Tree
The Barber Shop
Country Sales and a Lemonade Stand
The Opera House
The Backhouse
Characters in a Small Town
The Skeleton
Always In-between
Almost Never First
Forks in the Road
Several Careers
Out Where the West Begins

A Boy in Ohio

One does not write about a boyhood in the eastern part of the Midwest unless one has lived there. Holmes County lies across the western edge of the Appalachian Plateau where it merges raggedly into the brown soil-plains and lowlands of the true Midwest. Corn and wheat fields, forests, oil wells, coal mines, orchards, farms, creeks, streams, and the Killbuck River Valley typify this hill country of Ohio. There are old gnarled apple trees planted by Johnny Appleseed as he walked along munching apples and planting the seeds in the rich brown soils on the slight rises of the valley bottoms filled with glacial outwash and river sediments. Such was the countryside where I was born at five o'clock on a snowy, wintry Wednesday morning, the 17th of February 1915. The Berlin village doctor, Stremfer was his name, was brought to our home by Dad in his two-cylinder Maxwell with the side curtains fastened tight against the blowing snow. Of course this second son miracle was started nine months before, more or less, by my mother and father, in the old house on 17 acres bought in 1907 on the north edge of Berlin village.

It is not my intention to bore you with mundane autobiographical details so much as it is to outline the environments of a boy growing up in Ohio during the difficult times beginning with World War I and on through that part of the terrible depression to 1932 when I finished high school. Berlin had then 275 souls, and the day I was born an old man died and was to be buried in the lonely country cemetery where my mother's parents now lie. Some said this was an omen, a new one for an old one. Many old-timers had such superstitions.

The largest and best kept farms in Holmes County were owned by the Pennsylvania Dutch Amish-Mennonites, descendants of the hardy pioneers who escaped from Switzerland and its religious persecution and economic hardships. They came to Pennsylvanian America at the beck and call of William Penn. In 1915 the Berlin area was peopled in part by Mennonites, not Amish, as well as a host of characters of English extraction, many of whom belonged to no church—or if they did or ever had, they wouldn't admit it. Most of the fancy people with English names lived in the county seat called Millersburg, a town with some 2500 people seven miles to the west. Millersburg was named after a Miller who was probably some distant relative of the Miller family on my mother's side. There were two banks, a lawyer or two, two barber shops, and enough bars to wet lots of people's whistles until the Volstead Act became a curse on the country. There were also several hardware stores, a clothing store, tinner shops, a rubber factory, several mom-and-pop grocery stores, many churches, the grand old red brick Opera House, and a hotel where all kinds of interesting things took place. Eventually, before the recession of 1921, the A.S. Wengerd Slate Company warehouse was established by my father on South Mad Anthony Street near the station of a railroad line out of Wooster, Ohio.

In Berlin, the Myers sisters owned a dry goods store that sold some groceries, yard goods, and had the school book and supply concession. Other growing businesses included the store owned by D.W. Miller, who, my mother maintained, could not be any kind of a Miller relative, and a restaurant with a barber chair owned by Gilbert Miller. A half mile to the west was a slaughter house owned by Harry Miller, and still farther west near Martin's Creek was a large farm owned by Mennonite Melvin Miller. All these were sons of Sam Miller, the bearded patriarch of an old Miller family whose descendants have scattered far and wide.

There was one boarding house, a large clapboard structure built and owned by Joe Kandel. He was the tall, spare, white-haired Mennonite who led singing in the Mennonite Church in Berlin. He'd ding his tuning fork, hawk a bit, spit to one side of a front pew, and lead the congregation of less than a hundred off tune through the old-time hymns printed in hymnals whose use was dictated by the Mennonite Conference in Pennsylvania. Sunday School was taught by Ralph Aling or "Bunny" John Miller from 9:30 to 10:30, and all of us Mennonite boys were required by our parents to go. Church was long-winded with preacher Calvin Mast, at times with Elmer Varnes, and later with Simon Sommers. Sometimes guest preachers spoke, such as the famous Dr. Sanford Yoder, President of Mennonite Goshen College in Indiana, or Bishop Clayton Derstein, the well-known Mennonite author and spell-binder from the Mennonite Conference in Kitchener, Ontario. At times preachers from the Martin's Creek Church, the bellwether of local central Ohio Mennonite Churches, would grace the pulpit of our Berlin Church, the plainest of all churches—no piano, no organ, no music other than the glorious singing of untrained but enthusiastic voices "making a joyful noise unto the Lord." The sermons were long and tedious, filled with fire and brimstone about sins of the flesh and other sins, with pointed instructions that women would be damned to hell if they had their hair bobbed or wore fancy clothes and jewelry; boys were warned not to wear neckties, impressed with instructions always to tell the truth and never to fight, "turn the other cheek," praise the Lord, and watch out or "you'll go to hell." The sermons were, to a boy, interminable, so I spent the time whispering jokes with my cousins, writing short essays, eyeing the girls, or dozing off with my eyes open.

My father was what you might call a pillar in the community and widely known in central and northern Ohio for his wide-ranging business interests, his travels, his story-telling, and his service to the community. He would never sit at the front of the church but always in the second or third pew on the right side reserved for the men and boys. More than once he observed that old Sam, the Miller sons, Joe Kandel and some other Sunday Christians, who forgot their Christian ways during the rest of the week, always sat in the front pews. The left side was always reserved for the women and girls. Lots of winking and flirting went on among the boys and girls. When the preacher, chosen by vote of the congregation, noticed such antics, he would launch into a harangue about the sins of sex, which always amazed me because all the preachers had large families; someone must have been doing something! Much of the preaching was outright raillery against heinous sin, how to dress, how to act—direct holdovers from the long preaching in Old Order Amish-Mennonite church meetings held in barns. These early Amish were forebears of many of the Mennonites who escaped to join our more liberal, yet still drastically conservative, Mennonite churches.

In those days, perhaps throughout much of America, there was blatant hypocrisy, even by Mennonite elders and preachers. It seemed to a boy growing up and observing adults that this had something to do with the 1921 recession which had harsh effects on our farm country and business people. When the Depression hit, outright dishonesty bloomed. For example, in 1933 a man named Miller, no relation I was assured, caused a bank to fail and was put in the penitentiary in Columbus, Ohio. Several of the town drunks were the most rabid Sunday Christians. Others cursed, drank, whored around in Canton, and never went to church—at least they weren't hypocrites.

From the age of 11 to the age of 17 when I left for college, there was hard work on the roof for my father and on the farm taking care of 2000 chickens for my mother. Odd jobs such as washing wool in Doughty Creek or planting ginseng in local woods for Ralph Aling and lathing houses with Paul Hummel for Jerry Slabaugh helped me earn spare money to build model airplanes. No work was easy, and there wasn't much money. A dollar a day washing wool, and 25 cents a bundle for lathing; if you worked fast you could nail up eight bundles a day. My $2.00 a day—an almost unheard of sum of money—was spent on books and airplane model materials ordered from fascinating catalogs of shops in Peru, Indiana. Other kids got to go to Long Lake for summer camp or joined the Boy or Girl Scouts. I'd watch them go off, laughing, on the buses and return two weeks later, nut brown, with all kinds of tales. My father said it was a waste of time; work was more important. However, to his credit, my father and mother took us traveling: to the Cleveland Air Races of 1927 through 1932; to Washington, D.C., to visit the Smithsonian Museums, to visit the Senate and House of Representatives, the Capitol, the Lincoln Memorial, and the Washington Monument; to Niagara Falls to visit the shredded wheat factory, the Niagara Falls Museum, and the great whirlpool; and many times to Pennsylvania to see my Uncle John, Father's brother, and his family, and Stephen (Steffy) and Elizabeth (Betsey), my father's parents. Trips then of 15 hours, now done in four! From the back side of Negro Mountain, high on the Appalachian Plateau, I had my first glimpses of the mountains, the folded Appalachians, ancient and forbidding, covered with vast forests. Something deeply ingrained from my Swiss background made me resolve to live someday, somewhere, in mountains.

Visiting Uncle John Wengerd and his family
Salisbury, Pennsylvania
1935
Uncle John is at left rear. I am at far right.

We went to Gettysburg, taking with us my grandparents; now old Betsey got to visit her sister, Amanda, who was married to Sam Farmwald, one of the richest Amish-Mennonites in the Lancaster area. Betsey's sister inherited all the Shetler family money, leaving Betsey destitute, all because she married Steffy whom the Shetlers maintained was a ne'er-do-well. My father told me more than once that he admired the affluent Shetlers greatly, more so than his own parents—one sad commentary on my father's priorities. On the 1923 trip by car to Lancaster and Gettysburg in our open, canvas-topped Maxwell touring car, my grandfather Steffy got all excited when we visited the locale where Lincoln gave his brief historic address after a three hour oration by Senator Edward Everitt, a speech seldom mentioned thereafter. There is an old family anecdote (which I cannot document) that, in 1863 at the age of nine, my grandfather heard Lincoln's speech at Gettysburg with his father Moses, who had returned to visit his old haunts in Pennsylvania despite the raging battles of the Civil War. This is hardly believable because my ancestors were Amish-Mennonites, conscientious objectors, who hid in the hills and refused to fight in the Revolution, Civil, or any other war!

It is in the ages from first memory to graduation from high school that a boy's character and personality are most deeply shaped by family. Impressed by the strength of my father and mother, their great love of travel, the speed and forcefulness with which they dispensed corrective discipline for misdoings, their insistence that their children go to Sunday School and church, their early talk about education and college, and their fairness, one could not help but develop a strong love of family. The way they treated their parents was an eye-opener when compared to the way some of my friends treated their aging parents. I could not help but look at my parents, and later my wife's parents, as people to emulate in many ways. Of course there were arguments; in fact, my father baited me to argue as heatedly as he did with my older brother. Some of the raucous disagreements at the breakfast table between my older brother and my father, and even between my father and mother, left me cold to intra-family squabbling, and in my embarrassment at such contained anger, I could not participate. But I learned from such encounters how to stop a quarrel.

One of the greatest influences in my young life came via a collection of 78 rpm records which a ladyfriend of my father gave to my mother. She played those records on the Victrola by the hour during the half year I had to stay out of first grade because of an undiagnosed illness. For some reason, my left leg was pulled up by muscle constriction, and I couldn't walk. Mother read to me many hours in English, a language to which I had been introduced by the redoubtable

Miss Engel in the first half year of the first grade at the old Berlin school. These two pastimes, listening to classical music on records and to my mother's musical voice as she read to me, had a profound influence on me. Her work in our home was often left undone. Such times of sickness were the major reason why I realized that I needed to get all the education I could absorb, so I could eventually do as I pleased. We were never in poverty, in fact never broke; we always had plenty to eat and could travel as we wished. We cared enough about money and its value, the difficulty of saving it and making it grow, that we husbanded resources, grew much of our own food on the farm, raised chickens, butchered hogs and smoked the meat, and my mother canned hundreds of quarts of fruits and vegetables out of our own orchard and gardens. Nothing was wasted. My mother ran the house on egg money from the thousands of chickens. Guess who took care of those chickens!

A respect for the relationship of money and work was drummed into us early. Stories my father and mother told me about poverty among our neighbors were well-learned lessons about the hazards of a spendthrift life. By the time I was ten, I had saved $10.63 in my bank, a bronze shepherd dog with a long tail and a slot in his back. Counting the pennies, nickels, dimes, and rare quarters by shaking them out of the slot did seem to make the amount grow larger. Into the bank went the coins I found while at play, including the coins from under the seat cushions of our one over-stuffed couch and from Dad's chair, as such coins would dribble out from his pockets. Not until I was grown did he admit that he secreted coins there from time to time to see my glee at finding them.

After all, my allowance was only 25 cents per week, and it cost that to go to the movie at the old Opera House and eat a chocolate sundae at the Rottman drug store in Millersburg each Saturday evening. So how was I to save much money? A way presented itself after I was given a single shot Stevens rifle for Christmas when I was eight. Remington rim-fire shorts cost 25 cents for a box of 50. I engineered a contract with Dad that paid me two cents for each mouse tail, five cents for each rat tail, and five cents for each sparrow head. I became an expert shot, and in two years the dog bank was full as I shot up the barn and the trees, scaring everyone in the neighborhood. Dad then introduced me to banking. We opened a savings account at the Adams Bank where Dad did all the banking for his contracting business. I met Helen Adams, the pretty banker who worked for old Mr. Dwight Carey who owned the bank. Saving was fun from then on, but four percent interest on $10.00 or more seemed entirely too slow to become rich enough to learn how to fly. So I insisted my father increase my allowance to at least 50 cents per week. He agreed, and I quickly learned the art of diplomatic negotiation!

When I was two, the extended family held the first Alexander E. Miller reunion on April 22, 1917. I remember my fright at the great magnesium flash made by the photographer. My grandfather Alexander and my grandmother Catherine were seated in chairs in the middle of the front row. There were 27 people at this first A. E. Miller reunion, and that photograph of all the ten children of those two redoubtable grandparents, now long gone, is one of the most prized of my possessions.

A.E. Miller family reunion
April 1917

A photograph of the second A. E. Miller reunion, held in the spring of 1928 when I was in the eighth grade, shows 51 family members in attendance. My stalwart and kindly grandmother is alone in front, grandpa having died two years before. The family was greatly enlarged, though Aunt Mandy was gone and so was Oliver Lenhart, my Aunt Annie's husband. Several group photographs later, when I was a junior in college, found this now annual reunion so large that the group filled one whole side of my mother's large garden, minus Grandmother Catherine. In 1990 the last child of A. E. and Catherine Miller, affectionately known as Katie Ann, died at the age of 96.

The annual reunions are now in the hands of great-grandchildren, grown up, married, with numerous progeny. The early days are gone, replaced by other days, as time is wont to do. Descendants to the fifth and sixth generations still venerate old Aleck and Catherine. They had founded a Mennonite family of farmers, tradesmen, school teachers, truck drivers, professors, roofers, musicians, preachers, school principals, and a few scoundrels. All are loved as family; because if one has nowhere to go, the family is always a place to belong—a last bulwark against a world where family is often denied and denigrated. For me, the runt in my family who finally grew up somewhat, this extended family was a foundation like that of many Americans who sprang out of the soil and a conservative religion. It is said that you can take a boy out of the Mennonite Church, but you can never take the Mennonite out of a boy grown to manhood.

Little boy with dog

My sister, Carol Ann,
(right) with cousins and friend
September 1939

Never a Dull Moment

Every country boy lives his own life in his own way as much as is possible under restrictions placed on him by older people: mother, father, older brother, parents of neighboring boys, grade school teachers, high school teachers, local farmers, and tradesmen in town. This recitation involves a small country town in Ohio among farms and forested hills that lie across the band of glacial moraines left by continental ice during at least four times of maximum glaciation in the past million or more years. The hills were partly shaped by ice. Some were created by slowed ice running out of the power to carry the load of glacial till as the ice front melted. Cols were cut by fingers of ice riding over low divides and down narrow valleys choking the stream valleys with outwash to create flat bottoms where great farms now lie. Across southern New York, northwesternmost Pennsylvania, northern Ohio, Indiana, Illinois, Iowa, and Nebraska, these great sheets of ice created an environment which became the bread basket of America.

What a country to explore. Never a dull moment! The last time I was ever bored was at the age of four, just before I learned to read but still couldn't speak English. Then I discovered the funny papers. Because I was a middle child, I spent much time in the forests by myself. When my teachers took us grade school kids on hikes, we learned the names of bushes and trees, about collecting cocoons of butterflies and giant moths and about collecting leaves and flowers to press. But we got our ears pulled for being ornery and teasing the girls who always seemed to ask dumb questions and get in the way. In the winter time we skated on Boyd's Pond and went sledding on Kill Devil Hill on the north side of our farm—a dangerous steep incline that caused you to slide into a frozen-over but usually deep creek if you missed the narrow bridge we'd built across it. That creek was a southern tributary which started as a spring at the back of our house and flowed north into Lion's Run, which itself drained into Doughty Creek northwest of the town of Berlin. A wonderful country over which to roam and grow up!

There were favorite toys and vehicles: a clay-headed doll with a pink dress when I was really little; a low wagon with wooden rods on the sides to hang onto as my older cousins and my brother would try to spill me off; a red scooter which helped strengthen my legs as the wind flowed through my hair, cut short by Mother—the sense of speed was amazing to a boy made moribund by sickness through the first four years of life. Later when I was nine, my older brother gave me his old bicycle, and one day out in our spacious garage shop he taught me how to ride it. The greatest thrill was to be able to ride a bicycle; then came a flat tire and, for a year, I forgot about the bicycle. However, on my tenth Christmas, Dad brought me a bright new red bicycle with big fat tires. Heaven was here and now, indeed. That bicycle extended my range up to seven miles, west to Millersburg, north to Winesburg and Trail and to Mt. Hope, and south to Charm near the place where Father had his bicycle broken up by an angry Amish bishop many years before. Hair raising rides down long hills on unpaved roads were a favorite activity.

With almost disastrous results, Dad had one of the hired hands try to teach me how to drive a car at 13. At age 15 I learned to drive a heavy 1927 Hudson Terraplane Coupe while Dad took his fancy 1924 Hudson Brougham to work. It had been my ambition to learn to fly before I learned to drive, but Dad disabused me of that notion by saying, "I'll never let you learn to fly. Airplanes aren't here to stay!" I learned driving on my own. I didn't know how to turn around so I'd race up a quarter of a mile to where Mr. Krieger kept his steam threshing engines and "back" down to our house. Finally, I got the nervy idea to drive the whole mile to Bunker Hill where two roads crossed, thereby making a big circle to come back home. When I was 16, I discovered a cute girl new to our school. I grandly took her to a Sunday School Literary in the Hudson coupe and later to a movie, but Mother made me take our maid along as a chaperone; how droll!

Other fascinating things kept happening to me as I was growing up. When I was eight, Dad bought me a Stevens Single Shot Rifle and taught me how to use it; the year before he had given me his old heavy gold Elgin watch, a real treat, which I still have. At eleven he gave me a single-shot Western Field 20 gauge shot gun, and that same fall I shot my first rabbit late one afternoon at the southern edge of the Pomerene woods east of our house. One of the greatest thrills of my life was to see that cottontail somersault when I hit him. I took it home and Dad helped me skin it so we could have it for supper. From that first rabbit on, every fall, until my sophomore year in college, two Catholic friends of my father—the president of Republic Steel in Canton and Republic Steel's top salesman—would come to stay at our house to go rabbit hunting in "The Plains," a dried up glacial lake bottom northwest of Berlin. The rabbit season started October 15 and the limit was 15 rabbits per day per person. Rabbits were the scourge of the farmers who had truck patches in the thick black loamy soils around those grassy "plains," so rabbit shooting was a necessary sport. On every first day, Dad would organize and direct a line of six to eight hunters to sweep through the plains driving rabbits ahead, shooting only ahead. By evening we'd all have our limits and over, of rabbits, for game wardens never bothered rabbit hunters. When we got home after those gang hunts, we'd have as many as a hundred rabbits to skin and gut. Those millionaire friends of my father would store them in cracked ice and take every one of them back to Canton. My mother was the best rabbit cook around; wild rabbit was one of our favorite game animals.

When I was twelve, I talked Dad into buying me a model airplane kit, a Cecil Paoli twin pusher, which I put together with Dad's help. For a man who detested flying, I was amazed at his interest. But the instructions were sketchy and we didn't know whether to fly it forwards or backwards; in fact, we never did get that darn model to fly. One day after I had studied all the aeronautical books I could understand at the age of fourteen, I took it apart and used the parts to build a low-wing single pusher. It was my own design; it had no rudder, no dihedral, a large wing aft, and a small fore plane. This rubberband-powered airplane had a single propeller which I carved out of balsa. After several crashes I discovered dihedral and rebuilt the pusher. On a damp, muggy Ohio day in the hayfield across from our house, I launched the plane, and it flew! That was a thrill greater than riding my first bicycle or shooting my first rabbit. That successful flight launched me into a thrilling three-year career of building model airplanes.

My interest in science came about because of two teachers. The first was the stentorian Miss Engel—she of the high button shoes and the bun of chestnut hair on top of her head. She brought a pail of sand to school one day and put a pile of it into a basin, noting for us that sand always lies at an angle; that is, a pile of sand is conical with equal slopes all around; if you piled on more sand, the angle was always the same. I had to go to college to find out what it was called (the "angle of repose," to be exact). Miss Engel then poured water around the pile of sand in the basin and said, "Let's look to see what happens in an hour or so." Now, in the third grade, impatient kids expected some major catastrophe, perhaps an explosion or a lightning strike on the pile of sand. Nothing violent happened, but we kept watching the sand from our seats, to the detriment of study between classes which involved the six grades in one room. At the end of an arithmetic class for fourth graders, Miss Engel said, "Now let's look at what's happening." We all gathered around to find that the water had climbed uphill, wetting the whole pile of sand clear up to its conical top! Unheard of; everyone knows water can't go uphill. With that simple experiment, Miss Engel taught us the concepts of porosity, permeability, and capillarity, without using those big words which we had never heard.

The second teacher, Mr. James Miller (and he was always Mr. Miller), took us on long hiking trips when I was in the seventh grade. We looked at cocoons on leafless trees and bushes, at rocks with fossils and at grasses. He taught us in a local orchard how to graft fruit tree tendrils from one tree to another, how to identify trees by their leaves and their bark, how roots pull moisture up to leaves, about chlorophyll, and why leaves turn color and fall off. As a boy I wondered how one person could possibly know so much; it sounded like a lot of work! To learn all this stuff would certainly cut into play time and sports. But a die was set and I didn't realize it. Demonstrated intelligence was a matter of study, thought, and hard work—a tough lesson to learn for a boy who enjoyed recess more than study.

My very first scientific experiment was a real dud! One day after a heavy rain, some buff colored mud in puddles of wagon tracks had several thin hair-like black wiggly things that moved like very thin snakes. At age six, I discussed this with a friend who was a fourth grader. Convinced that he knew everything, I asked him what this could be. "Hair snakes," he said. I then reasoned, with what I much later learned was Socratic logic, that these were found in water-filled ruts made by wagons drawn by horses with black hair tails; therefore, if their black tail hairs fell into warm muddy water, then they must turn into hair snakes. With this bit of bright logic, I went to our barn where we had two horses, found some black horse-tail hairs and put them into warm muddy water after the next rain. I waited, I watched, and I waited several days and even went so far as to put more mud in water heated on Mom's wood stove. The experiment was a bust—no hair snakes. Much later, still disappointed at this failure of a perfectly logical experiment, I asked Mr. Miller, our science and arithmetic teacher, what those black wiggly things were. He called them "hair worms." Unsegmented and unique, they hatched from eggs spewed out by other hair worms. Lesson learned: the obvious is not always a good answer.

Whether it was true exploration or not, I spent a great deal of my time trying to find things: in the spring—generally April after the last snow—we searched for morels, which look like mushrooms but aren't. They grow overnight, to lengths of two to six inches with top-like white, buff brown, or black crowns on white stems. Our Pennsylvania Dutch name for them was michelin, mychlin or, more popularly, "feightly"—the moist ones. Once one found likely areas where these rarities grew, no one was ever told—competition was keen. Mother soaked them and fried them slowly in butter to make the most heavenly of wild food in the world. No mushroom is ever as good—a real treat.

On a hill north of our farm, where corn grew at times, lies an area where Indians made arrowheads. Each spring I found chips, scrapers, arrowheads of white chert, black flint, and even an axe head of stone—thrilling!

Fields of grain, wind-rippled as though bears were running through wheat or oats; woods with morels in the spring and squirrels to shoot in the fall; grassy areas near the corn field where rabbits hung out; animal burrows along our farm's creek where I trapped opossum, skunk, ferret, and, once in a while, rabbit in the fall; fields where bob-o-links, larks, red-wing black birds, and blue birds lived and sang; the chicken yard in whose dark damp soils I could always find fishing worms to go catch suckers and blue gills in Doughty Creek; the Doughty swimming hole, now long gone, where we swam after a hard day's work on the roof; leafless young trees along fence lines on our farm where I could find cocoons to collect and watch giant polyphymous moths emerge—all these led to unlimited joy and discovery for a country boy growing up in Ohio during the first third of the 20th century. It's true you can take the boy out of the country, but never the country out of a boy who becomes a man!

Dad Wengerd—Hunter

Nicknames

The Mexicans call a bald-headed man "Pelón," (the hairy one). We call some tall men "Shorty" or "High Pockets." Everyone should have a nickname; it's easier than a formal name. Nicknames don't stick to some people. I was called "Shermy" when I was little, "Lefty" when I pitched baseball, "Windy" when I was a roofer, "Wengy" by some of my faculty friends, "Wyoming" by a geophysicist friend who said Wyoming was the Sioux name for big wind, and "Shady" because I always sought shade on hot days, NOT for supposed illicit activities. Now it's "Sherm," which is about as unimaginative a nickname as one can have. Yet that's the one that's used. It's better than "hey you" used by some detractors, or "you there" if you're being called down. Of course, I really don't care what I'm called as long as I'm called in time for lunch!

The Amish-Mennonites are the prime jokers of our polyglot society. Their humor ranges from oblique to blunt to fittingly funny. Because there are so relatively few common last names (I think I counted about a hundred once), the Amish have a unique system of identification of common names. They tend to stick to biblical first names. This leads to many John Millers, Solomon Hochstetlers and David Yoders. It gets even worse when names like Christian Wenger, given to many of my ancestors, are repeated through the generations. How to solve this dilemma? The classic example is Louie Monie's Davey. This stellar individual was a fantastic man who died only a few years ago. He would have been a better baseball pitcher than Satchel Paige or Bob

Feller had he been allowed to pitch in the major leagues. But back to names. Davey's father was named Monie; his grandfather was Louie.

Nicknames usually derived from what people did or where they lived, or some unusual characteristic. Then we have "Broad Run John," who lived at a creek named Broad Run in Ohio; "Der Weis Mose" whose clothes were always dusted with flour because he was a miller; "Schmidt" Miller who was a blacksmith; and "Hussa Orsch John" whose back pants flap always came loose or he forgot to button up and would walk down the street on a hot day with his fanny exposed to the breeze. There was also "Hinkle Sheis" Solomon, a man who raised chickens and always had an odor of chicken shit about him. I had a distant cousin called "Fluch" Dan because he cursed so much. Another man was named "Berry" Dan because he raised strawberries. My father for a time had the nickname of "Sears Roebuck" Al because he railed at people who bought roofing through catalogs. That name finally settled down to "Al" until he became so well known that people called him "A.S.," for his real name was Allen Stephen.

There is another characteristic of Old Amish-Mennonite names that is rather unique. Many men are given their father's first name as a middle name. This died out in my father's generation after my grandfather Wengerd left the Low Order Amish Church because of a rift with a domineering Amish bishop.

Generally, the whole idea of a nickname is to identify a person more narrowly when many names are similar. Some of the names are longer because they describe the person. Other nicknames are derisive and even cruel. I have a friend whose nickname is "Tubby." If you called Noel Engel by his correct name, he'd look around to try to find the person to whom you were talking; in fact, he always introduced himself as "Tubby." He is a well-known geologist, now retired, who has become an award-winning and nationally-known photographer. Another friend was named "Toothpick" all his life because he was as thin as a toothpick. In many such names, endearment is meant. Women's first names are usually shortened so that Elizabeth becomes "Lizzie" or "Betsy;" yet Ann is always "Annie," Matilda is "Mattie," and so on and so on.

The nickname is, of course, not limited to Americans, but our rich polyglot language, so filled with idioms, is fertile ground for funny nicknames. To be incensed makes an unwanted nickname all the more applied. To be thin-skinned may even lead to worse names. In Mexico, a fellow I knew named Federico was called "Freddie," a nickname he detested because he was a tall, handsome, strong man. He protested so much that they called him "Freddie the Bandit," which is what he was. It pays to ride with the punches and accept a nickname as it is offered. A wise person who introduces himself with the nickname he likes makes for a smart man! The worst kind of ass is the person who says, "I'll bet you don't remember me!" There are answers to such stupidity, perhaps too raw to write down.

Berlin School

A Field Next to School

To the east of old Berlin School, built in the late 1890s, on the western edge of the Appalachian plateau of northeastern Ohio, lay a flat field. Such fields are a rarity in eastern Ohio, except along the valleys partly filled with glacial outwash from melting ice of the last glaciation. Located on a high hill, Berlin Village was first platted in 1816 and settled by English-speaking Americans until the early 1900s. It is the second highest village in the state of Ohio. Ted Sherlock lived in a neat house with a large red barn out back on the south side of Berlin's main street. His father had sold the site for the school to the local school board, which met in the township house to decide weighty matters. The village had no council, mayor nor manager, only the school board and a small committee which assigned grave sites in the cemetery on the northwest shoulder of Berlin hill.

The field south of Theophilus Sherlock's house and barn was adjacent to the school. "Top" or "Ted," as everyone called Mr. Sherlock (except us kids), allowed us to use the field as a softball diamond. By realigning and enlarging the base lines and building up the pitcher's mound a bit, a baseball diamond was created. Until about 1927, that field was the site of all kinds of community affairs, yearly festivals, games played by the Berlin town team, and one time, about 1925, the field hosted a one-ring circus. To us kids that was the event of the century. The circus had one lion, a tiger, some monkeys, a bear, a group of clowns, a small elephant, and a young giraffe. The circus travelled by truck. All excited, we watched the circus roughnecks, later to appear as clowns, put up the single tent, with its two poles and lots of guy ropes fastened to stakes driven in the ground by two strong men who hit the stakes alternately. We were fascinated by their rhythm; we'd never seen such a display of timing and strength. Later, these same muscular men in tiger skins were part of the show as strong men—to the "oohs" and "ahhs" of admiring teen-age girls and not a few dissatisfied wives of local men.

Cost of admission was 50 cents for adults, 25 cents for kids, but the hustling Beachy boys, Orin and Atlee, got in free by carrying buckets and buckets of water for the animals. A few kids of hill-billy families living in Millersburg crawled in under the tent flaps and got in free; a few were caught and tossed out, only to sneak in on the other side. The circus gave two performances, one in the afternoon which pre-empted the usual town team baseball game, the other in the evening with gasoline lanterns strung high in the rigging. The clowns were a riot, throwing balloons into the crowd sitting on backless bleachers, squirting water on kids. The two aerialists thrilled everyone. Kids had never seen such performances before. The lion tamer actually got in the cage with the lion and the tiger, snapping big whips and making them do all sorts of tricks. Would wonders never cease? A brave man in a cage with a lion and a tiger! Of course we had to have cotton candy and ice cream cones which were rare in the village of Berlin. Dad allowed me to go to both performances because I couldn't watch everything the first time around in this one ring wonderland, foreign to us farm boys. For some years after that circus in Berlin (the first and only time that I know of a circus coming to Holmes County), we reckoned time and dates as before the circus or after the circus. Circuses later came to Millersburg annually, and Ringling Brothers, Barnum and Bailey Circus came to Canton, Ohio, each year.

But things had to change. Theophilus Sherlock, a

spare serious man who never smiled, seemed burdened by a classical name of which people made fun; and he grew older, as we all do. One day about 1927 he put up a strong fence between the school grounds and his field and told our school principal, Jim Miller, my mother's cousin, that "You can't use the field anymore. I'm going to plow that pasture under and grow corn." Crestfallen, we moved the baseball field to the north on the other side of the Berlin Cemetery, and the village of Berlin lost some of its morale and zip. We never knew why "Top" did that. Perhaps because the school wouldn't buy the field at his price; possibly he was bitter because no one offered to pay him for use of the field; or maybe he just became old and crotchety after his wife died, as he himself did in 1934. So ended the use of a field which was really our only park, except for the beautifully-kept cemetery where my parents and my older brother are now buried.

Today, the baseball field is again in existence and is owned by Berlin School which, housed in a new brick building, became a complete grade and high school in 1929-30. Later, the school was expanded both east and west and converted to a very large grade school. The new Hiland School, the high school for all of eastern Holmes County, was built between Berlin and Walnut Creek to the east. Today, Sherlock Field is home to the new post office and a large grocery and hardware store. Progress marches on from the simple to the complex!

My birthplace and boyhood home

The Standing Tree

There have been many favorite trees in my youthful wanderings around Ohio. Nothing is quite as inviting as the well-spaced limbs of a tall tree. Alongside our vegetable patch stood six giant sweet red cherry trees. They supplied high places from which to look down upon Amish buggies travelling the road to and from Berlin. Those giant trees provided succulent cherries which I picked by the gallon to sell to neighbors. Of course, it seemed I ate more than I picked which led to an amusing incident of inadvertent evacuation when I was four years old as my older brother picked me up and tossed me in the air. Those cherry trees, long ago cut down when the highway was widened, now exist as cherry wood furniture. Mike Kline, who ran the sawmill out toward Winesburg, cut them down for finish lumber when I was away at college—a great loss during bygone childhood.

In front of our yard stood three giant maple trees. They were handy for climbing, cutting out "Y" shaped limbs to make sling shots, and even for tapping them one year during the 1921 recession to make maple syrup. Another cherry tree, with big yellow sweet cherries, stood in our front yard. A high straight horizontal limb held a rope swing where I learned, at the age of two, the thrill of swinging. Later we put up a tire on a rope, which led to whirling twists that made us kids appear to be drunk as skunks to the amusement of Mother.

By the south branch of Lion's Run, a creek that flowed north from a spring south of our house, was a stand of young elm trees. During the summer of 1925 when I was ten, I worked for Ralph Aling on Doughty Creek for one dollar a day washing sheep wool. Ralph, some 20 years older than I, had a youthful streak in him. One day he showed me a new aerialist trick. He climbed to the very top of a beech tree whose trunk was about five inches in diameter at the base. Grasping the very top, he swung out and the tree bent double, bringing him gracefully to the ground. Aha, a new game! I tried it on a smaller tree which was standing on the edge of the creek where we often went swimming and out of which I caught many bluegills and chubs for my mother to fry. As I climbed to the top, this tree decided to bend over the creek and dumped me in the water. I can still hear Ralph laughing as, sodden, I climbed out of the creek. When I got home, I decided to try climbing one of the young elm trees near our house. The tree stood over a wooden platform left from a farm building that had been torn down. Shinnying up the tree, I got up about 20 feet and, Aling-style, swung out. The tree bent double and broke while I clung to it, dumping me on my head on the wooden platform; in shock and in a daze, I sat up, looked around and noticed I had fallen among some projecting 12-penny spikes. Had I landed on my head on one of those, "Katie kiss the kumquat good-bye." I would not be here to write this. That ended my aerialist activities pronto.

The tree I really hated most was a peach tree past the grape arbor in our backyard. About ten feet high, that tree grew lovely yellow-meat peaches. But it grew something else: whips. Infractions were dealt with by my mother in a direct, forthright, and painful way. "You stay here!" she would order. Grabbing a butcher knife, she'd march out to the peach tree, cut a whip, and march back into the house with determination befitting Caligula. "Bend over," she'd say; and I certainly did, for she was a large, strong woman. These whippings were frequent enough that I swear she almost denuded all the small peach-bearing branches on that tree by the time I was seven years old. If you've ever been whipped by a peach whip, however lightly, you never forgot it—and the infraction for which you were punished was never repeated.

The trees one never climbed because of burrs, but which held the greatest interest for marauding boys, were the chestnut trees. They supplied not only chestnuts but also the fun of swiping chestnuts from irate farmers—but that's another story! In the hardwood forests near our home in Ohio, many hunting experiences came my way, including shooting squirrels and cottontail rabbits for meat; trapping ferrets, opossums, raccoons, and the skunks that lived in great holes under large trees; and trying to shoot crows that were always smarter than any boy with a rifle. I also once hunted red-headed woodpeckers so Lloyd Zuercher's dad could mount a pair for me, and it was common practice to carve our initials on large beech trees by Lion's Run.

The tree that I admired most of all was a very old apple tree that stood north of our barn and west of the small creek at the back of our barn. Full boled and never sprayed, year after year it grew the sweetest apples I ever tasted. Ripe apples from that tree were a real treat. Mom told me that old gnarled tree, now long gone, was planted by Johnny Appleseed many, many years before. I have no idea how she knew that. Trees were friends to an active boy like me until I was 11, when Dad made me go to work on roofing crews. Childhood seemed to have been very short!

Me with Wengerd cousins, Marion and Logan

The Chestnut Tree

They stood like lonely sentinels in pastures that were seas of grass. They had large leaves, prickly burrs, coarse-ribbed trunks and were full-boled. In the 1930s these trees were thrown one by one to the ground by God Himself. Their leaves were so distinct one could tell they were chestnut trees from a mile away. They were an attraction to little boys like a magnet attracts washers or the tin coins so many countries now call money. From whence had come such treasured trees? They were not planted; they just grew, always alone, one to a field, not always in the middle, but just far enough away from a field-enclosing fence to give a farmer time to get a shotgun loaded with buckshot with which to scare away acquisitive boys bent on stealing chestnuts.

The chestnut trees harbored birds, especially larks, and squirrels that, like hungry boys, coveted the juicy nut which was so difficult to break out of its porcupine shell. Given time, the prickly-spined shells would burst open of their own accord, but who had time to wait? A boy with a chestnut gleam in his eye is one of the most impatient critters in the world. Now I'm not talking about buckeyes, that large ox-eyed nut which looks like a chestnut but is so bitter nothing but ants will eat it when the shell rots off as the emerging rootlet tries to find a bed to grow in. No, I'm talking about the smaller nut—the great American chestnut.

Looking for areas of fertile soil in their westward trek throughout two centuries from Lancaster, Pennsylvania, through Ohio, Indiana, Illinois, and Iowa, early Pennsylvania Dutchmen watched for two trees. The chestnut tree and the wild black walnut tree told these best of all farmers that "This is the place!" Dark, rich, well-aerated loam on hillsides covered with weathered glacial till in broad bands across the bread basket of America. These thick, spongy soils retain the fallen rain in reserves high above the ground water table which feeds the springs in the valley, where it is available to the trees planted with them. At every early-found spring, with a chestnut tree or a walnut tree on the hillside above, there was a place for a home, barn, outhouse, and tool sheds for these farmers who used horses to help them till the soil and harvest the crops.

Very early on, these first settlers cruised the forests, alert for Indians bent on thwarting their mission to settle the land. Many an arrow from a sprung bow whistled through the air and found its mark, as these new people with the black hats, beards and white skin ranged through the land looking for places to farm and to raise their families. One of my ancestors, Jacob Hochstetler, was killed along with his wife in 1757 near Shartlesville, Berks County, Pennsylvania, by Indians who objected to his tilling the soil of their former forests. Two of his sons were captured, lived with the Indians and became like Indians before they were rescued. This gave rise to legends that some of the dark haired, black-eyed beauties, like my mother, had Indian ancestors thanks to the famous Hochstetler massacre—such things cannot be proven, although there are records of a great-great grandmother in the Yoder family, cousins of my family, having been an Indian girl.

But I was talking about chestnut trees, not Indians, though Indians treasured the chestnut as we did from the time my ancestors arrived in the 1700s. On a fateful day in 1931, that terrible year, a chestnut blight swept the land and killed every chestnut tree, so that by 1933 all were gone. Was this blight in retribution for my ancestors having taken over these fertile lands from the Indians? The wood was not wasted as Amish and Mennonite artisans created sturdy chestnut furniture now treasured as our family heirlooms.

Without my knowledge, a great thing happened; how many years back I do not know. In 1991 while visiting the graves of my grandfather, Alexander E. Miller, and my grandmother, Catherine Miller, I found three—not one—THREE large American chestnut trees growing around their graves. The cemetery stands near South Bunker Hill school which my mother laughingly called Bunker Hill College. She had attended that one room wooden school for six grades. There were no high schools within 20 miles then, back in 1893 to 1910.

When I found these trees in that cemetery, I sensed that the Indians may have forgiven my ancestors. My wife and I gathered many pounds of this most heavenly nut for drying to take to our children who never had the thrill of eating a chestnut, had never even seen a chestnut tree, and had never been chased away from a chestnut tree by an angry farmer who, for some strange reason, didn't realize that God put chestnut trees onto the hills of Ohio for boys to pilfer.

Harry Hinkle—Barber

The Barber Shop

As a lad of four my mother put a soup bowl on my head to get an even haircut. Therefore, my first trip to a barber shop sometime later was a somewhat traumatic experience. Mom said to Dad, "Take him to Millersburg and get him a haircut. I'm tired of cutting his hair because he won't sit still!" The barber shop had four chairs. The boss was a barber named Pete Bilderback who sat me on a board across the chair's arms and used a hand clipper. Mom's technique pulled enough hair, but these handclippers were real punishment. Dad sat by and watched, for he had to stay and pay the 25 cent cost of the haircut. Then I saw a man come into the shop, the likes of whom I had never seen; no one like that lived in our town or county. He was black! In Pennsylvania Dutch, a German Swiss dialect still spoken all through the Pennsylvania Dutch country of Ohio, I asked Dad, "*Verwas dut er net sie g'sicht wasche*?" (Why doesn't he wash his face?) This brought gales of laughter from all in the shop, and Dad explained, "He is a Negro and they have black skin." My first haircut in a barber shop, hurtful as it was, was a real success, and my mother was pleased. Also I learned that there were black people in the world.

Since then, many scores of years later, I still marvel at the modern American barber shop. Until I went to college, several barber shops opened in Millersburg, but my favorite for years was in the basement of our old bank building—cool, great odors of cologne, two chairs and the barber, Corley Close, a World War I veteran who had been badly gassed by the Germans at Verdun. He was a dapper, well-dressed man who would go into a back room for a minute or two during a haircut to allay the pain in his lungs. He was a jolly alcoholic. The second chair was manned by a friendly but sallow youth who had tuberculosis. From these two barbers I got an education on thc backgrounds of many people in my hometown, for they both loved to gossip. Corliss "Corley" Close shot himself years ago, but both he and his helper are remembered as having been, along with Pete Bilderback, the best barbers in the business. Years later, Bob Casey cut hair for a long time; I knew him well. Today, his son, Bob Jr., a big man, former mayor of the town, is my favorite barber. I return to him for a haircut every year. He is my ideal source for gossip on everyone and everything in Millersburg, plus he is a great storyteller. Earlier, one of my cousins was a barber in the village of Berlin, but he got tired and quit.

In Texas, Wichita Falls to be exact, my favorite barber was an older man named Peavey, a true philosopher. In 1946 haircuts were still 50 cents, and when he had to raise prices to 75 cents, he was deeply apologetic to all his many faithful customers. On arrival in Albuquerque, the Campus Barber Shop on Central Avenue opposite the University was the place to go. It contained four chairs and was run by a talkative man—affable, kindly and an artisan with the electric clippers. Haircuts were $1.50. But he retired and moved to El Paso, so others took over.

About 1956, a young man appeared on the scene to help a retired sailor, "Salty," cut hair in the Parkview Barber Shop. Then "Salty" moved away and the new barber took over. Haircuts were $2.00. I noticed his name—Harry Hinkle—on a plaque above his mirror. Astounded, I told him a man named Harry "Pete" Hinkle worked on and off for Dad in the roofing business for many years. The barber said, "He was my uncle. I used to come up from Kentucky to visit him. I knew your dad and brother. In fact, I remember you when I used to play in your dad's warehouse." Then a long forgotten incident struck me. This was the pugnacious boy who once punched me in the nose because Dad told him to go home and stay out of the warehouse! So here in Albuquerque, 31 years after our boyhood fisticuffs, we became friends. Harry was the biggest liar I ever met. He won the "Biggest Liar in New Mexico" contest one year, defeating the former champion, Professor Frank Hibben, a faculty friend of mine at the university. Harry let me know that he was a retired lieutenant colonel in the Air Force who flew over Germany in World War II. Actually, he was a sergeant early in the Korean War and never went overseas! He could tell lies so adroitly that he made life interesting though considerably skewed. Some years ago, he gave me a haircut just a week before he died of intestinal cancer—a severe loss, for he was a brilliant conversationalist and a friend. Haircuts were by then $3.00, no tips expected.

Today I get haircuts at Kirtland Air Force Base, having abandoned the overly expensive Parkview Barber Shop shortly after Harry died; the two Spanish-Americans boosted the price to $6.00 for a haircut that takes only 15 minutes.

I've had haircuts by lady barbers in New Zealand and in Norfolk, Virginia. They're good—great talkers. One of the most interesting haircuts I ever got was in a shed on a high backless chair in Baja California. The barber reeked of Tequila, and pigs and chickens walked in and out of the shed—and most of the hair went down my back. But the cost was only eight pesos—64 cents!

Just recently I got fantastic haircuts from a Bangladeshi crew member named Abdul Barick on board the *Ivybank*, a freighter bound from Antwerp through the Panama Canal to Tahiti. He was another artisan with only scissors and comb! I estimate that I've had no less than a thousand haircuts so far in my life time—on board ships, in jungles, on a sailing vessel in the Arctic, on every continent except Antarctica (which I've visited but couldn't find a barber shop), on atolls in the Pacific, in Angola, Africa. I've gotten my hair cut on the North Slope of Alaska by an Esquimo hair chopper, in Central America, in South America, in Canada, in most European countries, on Army Air Force bases in northernmost Quebec, on Baffin Island, and in many states of the U.S.A.

A haircut is a very personal thing—therapeutic, even a sensual experience. It also involves a matter of trust, especially in some places where straight razors are very sharp and handy. It doesn't do to insult a barber; they have a code about unruly customers. I have heard of one loud mouth who almost got his ear shaved off for propositioning a manicurist who was working on his nails while he was getting a haircut. She was the barber's wife! The old saw is still told about a black barber slicing the throat of a "Whitey" and challenging him to try to turn his head. In foreign areas, barbers always ask where one is from. This leads to all kinds of language goofs, especially in France or Italy, two languages with which I have less than a fluent speaking knowledge. But even there, their solicitous demeanor leads to knowledge not gainable in any other way. In a strange city, if you want information ask a librarian, a policeman, or a barber, but ask the barber first! All are fonts of knowledge. In fact, my education would not be nearly as complete as it is, if I'd never gotten a haircut. Also, my hair might be rather long!

Three things pain a barber: seeing long hair on a man, having to give a "butch" (the typical military haircut), or knowing that a woman cuts her husband's hair. But strange things are happening. There are towns where the jovial, informative old-time barber is being replaced by suave young men and fancy women who run hair salons where the cost, just by the name, is double what a good old-fashioned haircut should be. My hat's off to a good barber! Good thing, because it's difficult to get a haircut with a hat on.

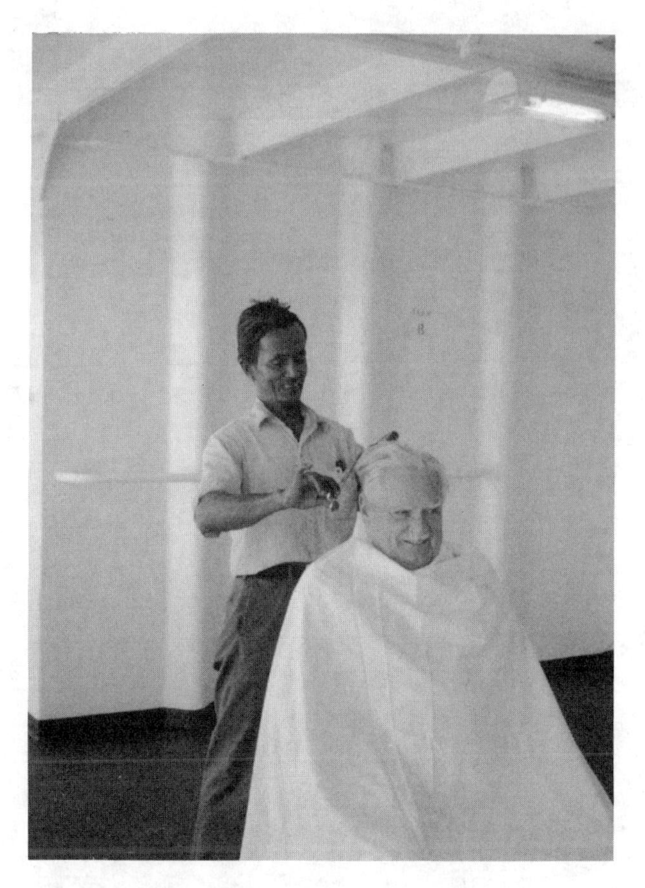

A haircut aboard the freighter *Ivybank*

A grown-up Paul Hummel
and his wife, Mary

Country Sales and a Lemonade Stand

For many years, Old Sam Miller promoted and ran the Berlin livestock sales on the first Friday of each month during spring, summer, and fall. He hired Levi Yakley as auctioneer at percentage of each sale, but Sam pocketed a larger percentage from rental of the sales ground. The sales ground had pens for cattle, sheep, pigs, horses, and, at times, goats, plus a raised auctioneer stand, covered from hot sun, rain and the rare snows of early spring and late fall in the 1920s and 30s.

These sales brought many hundreds of farmers, tradesmen, cattle buyers, and cattle sellers to the village which stood second highest in elevation of all villages, towns, and cities in the state of Ohio. Business was brisk from early morning to night. Now and again a bit of wagering went on—and, of course, the wetting of whistles out of bottles in paper bags, for Berlin was a village without a bar.

On the sales grounds, some four acres in size one block north of Main Street, stood a refreshment stand manned by Old Sam's son, Gilbert, who owned a restaurant on the north side of Main Street. Here, Gilbert sold pop, sandwiches, candy, and ice cream at rather high prices.

A high school buddy, Paul Hummel, son of the town butcher, and I saw an opportunity to make some money to further our own hobbies, so we approached a mechanic who had a garage on the west side of the street right outside the gate of the sales grounds. One whole summer, early on each sale day, we built a counter covered by cardboard and canvas in front of the garage. We borrowed crocks, ladles, measuring spoons, knives, and squeezers from our parents, and we went into the business of making the best lemonade ever sold in the state of Ohio. With oranges and lemons acquired from D.W. Miller's grocery store, we made tens of gallons of lemonade, floating slices of orange and lemon in the lemonade in a large glass tankard standing on one end of the counter. At 10 cents a glass, we sold many hundreds of glasses of lemonade each sale day. Our stand was mobbed by thirsty farmers, Amishmen, Mennonites, and fancy people from cities as far away as Cleveland, Canton, Akron, and Columbus who came to these sales. The auctioneer made a bundle each month, for this was the largest livestock sale in northern Ohio, known far and wide as the place where the finest farm animals raised by Amish and Mennonites could be bought. Of course, some scrub stock was brought in by ringers from outside farms in places like Loudonville, Dover, New Philadelphia, and Massillon.

However, the man who made the most money, other than farmers selling prize stock, was Old Sam, already the richest man in the village. Business at Gilbert's refreshment stand plummeted. "Why should we buy from Gilbert," asked the thirsty crowd, "when we can buy such good lemonade by walking just a few yards outside the sales ground?" We joked with our ever-thirsty patrons in Pennsylvania Dutch as they watched us make the lemonade on the spot, with chunks of ice in the glass tankard. For eight hours each Friday, we were busier than two one-armed paper hangers. After the sale ended, we went to the grocery store to pay for the lemons, oranges and sugar which we had put "on tick," as they used to say, and then we divided the money equally. Now, D.W. Miller's son rented that store building from Old Sam, who owned much of the commercial property in town, and he was privy to our success. It was mused about town how much money we were making because as the summer went on, Paul and I would EACH make up to $10.00 per Friday—quite a bit better than 25 cents per bundle nailing up lath for Jerry Slabaugh, a dollar a day washing wool for Ralph Aling, or shooting sparrows at five cents each. We were rich in the days of the Great Depression. Of course, we disassembled our lean-to lemonade stand after each sale day and thanked the mechanic for letting us set up in front of his garage.

This went on for the whole summer. At the beginning of the second summer, we arrived on the scene with hammer, nails, saw, and our knock-down bench and counter with the cardboard roof ready to put up our stand. The mechanic came out and said, "I can't let you sell lemonade in front of the garage anymore!" He had admired our early teen-age entrepreneurship, and, of course, we supplied him with free lemonade while he worked in the garage during the sales. We cajoled, tried to find out why we weren't permitted to sell lemonade, and even set up a stand out of the backend of Dad's terraplane coupe on Main Street, a block away. But it was no good; the stand was too far away from the sales ground for people to visit while important stock was being auctioned.

Crestfallen, we researched that refusal. We found that the owner of the garage building forbade the leasing mechanic to let us use the space, despite the fact it was on a public alley. Our sales of lemonade were driving down the income at Gilbert's stand—a part of which Gilbert had to pay to his own father for rental of space on the sales ground.

Through our experiences as makers and sellers of lemonade, Paul Hummel and I learned about monopoly, cut-throat tactics in business, mean-spirited rich men, profit-and-loss, and a great deal about people. Old Sam has been gone some 50 years now. Paul Hummel became a wealthy business man and a Mennonite preacher. I did okay as an independent geologist and consultant in oil and gas exploration out West. At times of remembrance today, people still talk of Paul and Sherman selling lemonade!

The Opera House in Millersburg

The Opera House

It was an imposing structure that could have been a castle in England or a fortress in Germany. Until an officious town council decided that it was ugly, out-of-place and useless, this great pile of red brick served as a catch-all for town offices that would not fit or which had been banished from the stately county courthouse in Millersburg, the Holmes County seat. On the sort of mezzanine level was a theater of imposing dimensions. In the late 19th century and early 20th century, touring musical groups performed operas and plays on its wide stage. All parts of the stage could be seen from every seat in the orchestra—from the loge back under the balcony and from the semicircular balcony high above the stage. An impressive house it was.

The opera house was a magical place with an imposing turreted tower, arched cupolas and windows, buttresses, and even some strange minor gargoyles. Parts of this vast structure were roofed with Vermont slate, and the tower had a green copper roof. This vast edifice stood diagonally northeast of the courthouse across the main street of the town. To the ultra-religious it was a den of sin, for theatrical productions in this predominantly Amish and Mennonite county were works of the devil. The front steps led up to a foyer with a ticket window. Wide curving steps made their way up to the spacious balcony; doors beneath the balcony steps led into the lower part of the theater. All inside was quiet; one got a sense of grandeur felt in no other place closer than Loews Theater in Canton.

The grandest use to which this opera house was put

was as a movie theater after 1920. That movie theater was where I got much of my education about the fancy people, as we Pennsylvania Dutch boys called the Catholics and English- speaking Protestants who peopled most of the large, fancy houses of Millersburg. An earlier movie house in a small building on a side street was the place where I saw, as a small boy of five, the first showings of a silent film about World War I and our doughboys being killed by Germans. A local man narrated the film taken by Army photographers. Perhaps it was Corliss Close, the barber, because he was there and was wounded.

The theater in our opera house became very popular as movie audiences outgrew the small theater on the side street. "Beans" Bowers, a tall, slender chap, became the projector operator. There was only one projector, so while "Beans" was changing reels, lights were turned on and a colorful advertising screen was lowered. There was always a piano player to add drama to the silent films—classics like *The Ten Commandments*, *The Great Train Robbery*, *The Third Alarm*, *The Great Train Wreck*, *Message to Garcia*, and *The Orphan*. There were also *The Little Match Girl*, *Ben Hur*, *Robin Hood* with Douglas Fairbanks, *Thief of Baghdad*, and *Son of Zorro*. Cowboy actors thrilled us boys with their derring-do—men such as William S. Hart, Tom Mix, Ken Maynard, and Hoot Gibson. Great actors like Ralph Lewis, Monte Blue, Francis X. Bushman, Frederic March, and others were favorites. And who can ever forget Lon Chaney in *The Hunchback of Notre Dame*? The comedians added much to our famous but obtuse Pennsylvania Dutch humor, and we would mimic them in school the next week. We could never forget the antics of men such as Buster Keaton, Harold Lloyd, Ben Turpin, Charlie Chaplin, Fatty Arbuckle, Laurel and Hardy, and Joe E. Brown. There simply aren't comedians like those anymore unless you include such classical later ones as

Danny Kaye and Red Skelton. They all made life extremely interesting, and we spent hours discussing their jokes and pratfalls.

Saturday night was THE night to go to the movies. Matinees were rare. I looked forward to Saturday just as I looked forward to recess at school. My allowance for years, it seemed, was 25 cents per week. Dad would say, "Don't spend it all in one place." The day involved paying 15 cents for the movie and then a trip to Rottman Drug Store to look over and read magazines, before buying an ice cream sundae with real chocolate sauce or a chocolate soda for a dime. The routine was ingrained; I went to the movie, Mom shopped, and Dad told stories on the street, for he was the county's greatest storyteller.

The opera house was an educational institution—a place where, along with school, I learned to read English. Many of the people in these early silent movies were beautiful, but some were bad and we booed the villains heartily. The piano player banged out racy tunes while cowboys chased Indians. The cowboy in the white hat always won the girl, while the cowboys in the black hats were always shot dead, to our loud cheers. We learned of death, pathos, exaggerated emotions, joy, hilarity, humor, and life—all vital parts of a boy's education.

The opera house is gone, torn down in 1954. The space now houses a dull parking lot. Such a loss should not have been allowed, for the building was solidly constructed and could have been made into a classical museum. Early politicians in Holmes County were not known for their historical sense. Given time, I hope to research the actual background of this classic structure, find out who designed its grand lines, who authorized it in an earlier day, and who built it in 1886. Then I'd like to find the asses who had it torn down. Most likely that's not possible, for death has undoubtedly claimed those numbskulls!

The Backhouse

The Backhouse

In the old days people ate Sunday dinner indoors and went to the privy out back to do what one ordinarily does in a privy. The little house out back, often called the outhouse, had a critical function in one's continued well-being. The backhouse was also called a Chic Sales, after a comedian who made a name for himself in show business by extolling the virtues of the privy. This little building was the scene of many adventures for which it was not designed. Boys and girls early on learned about sex away from the watchful eyes of neighbors and parents. Many learned to read by perusing well-illustrated catalogs which had been relegated to use as toilet paper. Before the widespread use of catalogs, newspapers were used. Even earlier on farms, corn cobs (without the corn naturally) and corn husks served the purpose but didn't supply reading material as catalogs did. Now, with modern indoor toilets, wags will note that Sunday dinner is often served outdoors as a picnic, and people go indoors to the toilet!

The old Berlin school was a large roomy two-story structure in a "T" shape. It had two rooms on the ground floor—one for grades one through six and the other for grades seven and eight. Across the top was a long room with a chemistry closet for the three high school classes. Out back stood two large four-holers which served all eleven grades of pupils. This mix of boys and girls utilizing two four-holers out back led to many humorous events. Knot holes were knocked out to yield peep holes. Shrieks by girls were common. Girls would pour water over the wall behind which was the boys' piss trough—OK, urinal, if you must be fancy. Boys would have contests to see how high they could piss over that wall—especially if girls stood outside to pour water on the boys inside. After all, there was no radio, no TV, no local movie theater in our small country town. What were boys and girls to do for fun, other than games at recess, such as blind-man's bluff, rounders, andy over, prisoner base, and sock or dodge ball?

At our house we didn't have indoor toilets until 1918. My father was always ahead of neighbors in adding new touches to good living. He had our plain oblong house rebuilt into a classic bungalow. He introduced an electrical generator, running water from our well, two toilets, one wash basin, and a bathtub. But the old privy, called "the backhouse" by all Pennsylvania Dutch, was not abandoned until 1950. My ancient grandma Betsy and grandpa Steffy lived with us in 1920 and 1921, and again she lived with us until her death in 1948, 23 years after Grandpa died. She always used the backhouse until she died because she was afraid the flush toilet would do her in and flush her away.

It was a common stunt in the days when everyone had a privy for Halloweeners to push over outhouses, especially those of unpopular families such as the rich, the store keeper who wouldn't extend credit, the tight-fisted business man, and other big shots in the community. "Berry" Dan Miller was a wealthy Amish farmer in Holmes County, rich enough to have a two-holer with both a star and a crescent moon cut into the door. The star was usually reserved for women's outdoor toilets and the crescent moon for men's toilets. One evening at Halloween time, "Berry" Dan was sitting in his fancy privy when suddenly it toppled over onto the door. Cursing the wind, he thought, "I'll holler loud so my boys can get me out." No one could hear him so he had to tear a hole in the roof to climb out. There were several jokers in the neighborhood—the kind who string ropes across town streets to stop Halloween buggies and rare automobiles. One such was Miller Dietz, who, with some of his cronies, had upset "Berry" Dan's two-holer and stood on it until Dan's head popped out of the hole in the now vertical roof. For awhile, Pennsylvania Dutch humor dictated that "Berry" Dan be called "Backhouse" Dan, which embarrassed him no end.

The design of backhouses was highly individualistic. Some faced away from the house, so the door could be opened and the occupant could watch the scenery. Others had the door facing the house for rapid entry in case of diarrhetic emergency. Some were large ornate affairs, others a simple bar for a fanny to hang across. They were of all conceivable colors but usually a weather-beaten gray. For a building so often used by a large family, only a modicum of effort was made to abate the characteristic earthy odor. The usual chemical was lime, which was evil-smelling stuff and as bad as what came out of humans. Some had screens over the door design and a vent pipe for odors; others were open in the back below the seats. It was a chilling experience to use such a poorly designed privy in the winter, and in the summer, blue flies and hornets inhabited the area beneath the seats, as well as opportunistic mosquitoes.

With the door locked from the inside, the outside hook hung as a signal that the backhouse was occupied. Odes have been written about pleasures experienced in those early necessary buildings. When indoor plumbing became common, there passed from the American scene a little building where much learning took place. Time marches on. Who misses privies? No one!

Ralph Aling

Characters in a Small Town

Any person who lived for 17 years in the first third of the 20th century in a small town in the Pennsylvania Dutch country of Ohio knew many characters and heard really tall tales about them. It would take a book such as Sherwood Anderson's *Winesburg, Ohio*, and its rendering of the town of Clyde, Ohio, to do these stories justice; however, some outstanding ones deserve mention here. There was Steve Kaser, a huge, tall mountain of muscle, the strongest man in Holmes County. Tales of his strength still abound 60 years later. He was a timber worker. One rainy night he walked across the flood-swollen Doughty Creek carrying an Amish buggy with a woman and two children while the horse swam the creek that flooded the road—an amazing feat.

And the ancient German, Andy Forsch, was a wagon maker who escaped Germany just before Bismarck tried to impress him into the German Army in 1870 to fight in the Franco-Prussian War. He was bald headed, had a long drooping moustache and a voice like a bull horn, and all of us kids gave his house in Berlin a wide berth while walking to school. As we grew older, we realized he was a kindly man who just knew that America was the greatest country on earth. Another character was Mike Kline's grandfather, Benedict Kline, who could lift a full oaken barrel of cider and drink out of the bung hole. He heard that Abe Lincoln, whom he had once met in Illinois, could do the same, so Benedict thought, "If a skinny rail like ol' Abe can do that, so can I." And when Benedict was already an old man, I saw him do it.

A boy of fourteen is easily impressed by such strength!

Who can forget old Joe Kandel, who ran a boarding house where his wife did all the work. He led the singing in the Mennonite Church in Berlin. Though not always in tune, the congregation soon got it straight, and the hymns were tumultuous—if there's one thing Mennonites do well, it is to sing loud and clear. Joe never seemed to have a regular job, so when Dad needed a hayfield cut and hay stacked in the barn for our cow and two horses, Joe was available to help. Joe's wife, a wonderful woman, ran the boarding house, a long, tall clapboard building on the north side of Main Street, while Joe sat on the porch resting.

There were the brothers Jake and Bris Kaser who also always seemed to be available for odd jobs. Jake was an early hunting buddy of Dad's, and they made money shooting coon and possum, trapping skunks, and catching ferrets for their skins which they cured on specially shaped boards and sold in Canton. And there was Boss Yingling, a mysterious man some thought was part Indian. No one knew where he came from.

Ike Hummel, the butcher, came to Berlin in about 1926 from Saltillo, Ohio, to run the lone butcher shop. Saltillo is a wide spot in the road among the forested "sticks" of south-central Holmes County. Ike was a fine butcher who loved a drink more than now and again. Some said he never drew a sober breath; but he could sure cook up a storm in his restaurant next to the butcher shop. He was at times an angry man, but he and his long-suffering wife raised a son, Paul, who became a well-known Mennonite preacher. Ida Vogt, still alive in Trail, Ohio, is a forthright woman who feared no one and ran the best supplied store in Holmes County. Her father started the very store where my parents bought their first housewares and some furniture after they were married in 1906. Dad teased Ida, but she held her own and could curse like a trooper. She is a lady no one can forget.

And there were many other characters such as Herb Teal, Woody Myers, Warner Maxwell, Frank Mapel and his sister Salome, Miller Dietz, Honey Myers, Lincoln Yoder, the auctioneer Levi Yakley, the historian Eli Reidenbach, the farmer Jerry Slabaugh, Florence Reidenbach, Wally Hampsher, and the Maple sisters, two old maids who ran the old post office and were my wonderful contact with the outside world. There were also Alvin, Gilbert, Harry, Melvin and D.W. Miller, all successful businessmen and sons of Old Sam, the shrewd businessman who reputedly owned half the town. Other characters were Jim Miller, the brilliant teacher who became a clock maker; Esther and Olga Myers who ran Myers store where we bought all our school books and supplies; Fritz Landalt, the German immigrant who was called "The Millionaire Hobo"; the old hermit, who lived on "The Plains" and cut reeds to make baskets; Fred Brown, the coal miner; El McCullough; Fred Miller, brother to Jim, and both cousins of my mother; Levi Gardner; sly Gideon Helmut; Ralph Aling, the genial philosopher; Calvin Mast and Simon Sommers, the early Mennonite preachers; and Al Wengerd, my own father. Dozens more were the kind of individualists one can't find any longer in the area where I grew up. We learned much about human nature, the good, the bad, the dishonest, the Sunday Christian who'd cheat the shirt off your back during the week, the genial drunk, the strong men, the curser, the weak, the outcast relative, the storytellers, the pathological liars, the stupid, the intelligent, the itinerant preachers, the crafty businessmen, the laborers, and many who were easily forgotten. Every town between the Civil War and World War II had its own panoply of characters. More will yet be written about people of that historic period from the small town of Berlin, Ohio.

East Main Street
Berlin, Ohio

The Skeleton

Her hollow eyes look out on the world from a shelf in our bedroom, her jaws fastened together so her several teeth won't fall out on our bed. High cheek bones and beautiful high-domed forehead tell us she was intelligent, and from the shape of her whole head, I could envision her as a beautiful young woman—perhaps 34 years old—when she died in child birth over 150 years ago. Her limbs were straight, rib cage full, and vertebral column gently curved. Her skull is a dark brown from a fire that ravaged my office in 1952. The fire was set by an arsonist, a neighbor boy who found his way into our home one winter's night, stuffed my denim haversack on top of a gas heater whose pilot light had been left burning while we attended a scientific meeting in Los Angeles. I should keep her skull in my new office to remind me of my own mortality as I try to fathom the history of the earth and write down thoughts about life.

Sometime before 1840, a Midwestern doctor had attended Jefferson Medical School in St. Louis, one of the first medical schools in the Midwest. He was young, research-oriented and had a medical practice in a small Ohio town. One of his patients was a young pregnant Indian woman who worked as a maid. Her morning sickness drove her in fear to the doctor who found she wasn't married and had no family. When she died in still-birth, no one claimed her body, so the doctor claimed it from the local funeral home. By means not entirely clear, he obtained permission—or avoided requesting permission—to dispose of the body as the county had no funds for burial. The doctor remembered the dictum of St. Benedict: "It is easier to beg forgiveness than to seek permission." After all, articulated skeletons of indigent humans were necessary in medical practice. Avoiding the soubriquet of the "Ghost of Gladstone Glade or Gaylord Gulah," the doctor dismembered the body as he performed an autopsy. He rendered the remains with care so he could articulate the skeleton for his office to demonstrate bone structure to patients, many of whom suffered broken bones in the hardwood timber industry of early Ohio.

The scene changes to 1931 in Holmes County. A grandson of the doctor, Fred Reed, then a mature man in the dry cleaning business in Millersburg, had the now partial skeleton and told me some of the facts aforementioned. Some of the story was surmised by both of us. He asked me, "Would you like to have the skeleton? I know your father wants you to go to medical school." Because I rather halfheartedly thought of medical school while I was in high school and was vitally interested in how the human body functioned, I said, "Yes, I'll take it."

So began the saga of what remains of that Indian girl's skeleton and how I now have the brooding skull. I tried pre-med and took many of the necessary courses in biology, chemistry and even physics, but I found my interests to lie not with blood or sick people, but in the study of rocks. After all, the outdoors were appealing, and rocks, though they tell stories, don't talk back. So I studied geology.

I took the skeleton to college—what was left of it—and hung it in my room. The leg bones had fallen off, the hip became unjoined from the spinal column, the arm bones fell off, and the well-formed teeth loosened in their sockets. But while it was still a reasonably complete skeleton, a college chum of mine and I decided to pull a typical college student prank. After all, it was during the Depression; there wasn't much money to raise hell with, other than to have a beer on Friday nights after a week of serious study.

I had a Model A Ford with a rumble seat, a 1929 Sport convertible that old Henry Ford decided to make with a top that would not fold down. "Too expensive," said he. This was a second-hand car bought for several hundred dollars from a car lot whose owner told me the usual story: "It was owned by two old maid school teachers, and they only drove it 15,000 miles in three years." My college chum, who later did become a doctor, and I put the skeleton into the rumble seat of my Ford, tied it so it wouldn't fall over and taped an old snap-brim felt hat on the skull. We drove that skeleton slowly all over the campus and the city of Wooster for two hours on a Friday afternoon hoping to shock college administrators, faculty, students, and town people. Not one notice was taken! No police stopped us, no one raised a ruckus, no *Wooster Record* reporters interviewed us, nor did the student newspaper, *The Wooster Voice*, mention this humorous escapade. Downcast, we returned the skeleton to my room, realizing that no one gave a darn!

Today, I wonder about the Indian girl as I look at her lovely skull. She was about 5' 4" tall, but who was her lover? How pretty was she? Skeletons keep their secrets well!

My sister, Carol Ann, and me
September 1939

My brothers, Wilmer and Owen
c. 1960

Always In-between

Why is it that some people are always doomed to be in-between in their lives? Admittedly no one can control when he will be born into a family—just as we cannot know when we will die. As the second boy in our family, I was followed by a third boy four years later. My older brother was born just over eight years before I arrived on the scene. A sister didn't appear until 14 years after my little brother was born, too late to have an effect on my own feeling that being an in-between was precarious. My older brother was moderately tall early in life; my younger brother grew to be six feet tall with a huge hulk of a body. I was short, obviously the runt in the family. I had multiple childhood diseases, including pneumonia three times before I was five, and a mysterious disease which our doctor could not diagnose. I was only four feet tall at the age of nine when I should have been on the way to a height of five feet. Fortunately, I grew until I was twenty years of age, topping 5' 8".

Left-handed, short, spindly, I was put on roofing crews at the age of eleven to do THAT work for eight summers. This bolstered the vitality inherited from my father and his mother, and I had a chance to exert myself. I developed into a chunky lad with broad shoulders, heavy bones and strong muscles. At the age of 16, weighing 140 pounds, I could bench press 300 pounds yet was lightning-quick on the high school basketball team. The classes ahead of me in high school, and after mine, were filled with big bruisers with long arms that made boxing hazardous for a lad with small hands. Always in-between! I felt, perhaps without foundation, that everyone was a lot smarter. I learned to read English before I could speak it. Then a whole new world opened up. My father insisted I could do whatever I wanted to do, go wherever my ambition would drive me, and that I should never be lazy until I could afford to be lazy!

All through college, my true love was athletics, yet my athletic career was not outstanding. I could hike all day and not tire, but both my brothers could run faster and swim faster than I could. What is there about in-betweeness that is so debilitating?

The struggle took a new turn when I found that I had an analytical mind despite difficulty with mathematics. Three-dimensional imagery was easy, writing was fun; I even read dictionaries—a strange avocation for an outdoorsman. Reading came naturally because of some driving force that made me want to learn everything possible about my world. While other boys were fighting or chasing girls, I was studying aeronautics and building model airplanes. Did all this happen because Mother read to me hours on end while I lay ill? We never knew what caused me to miss the second half of my first year in grade school. Many years later, after my Naval career was over, I tried to find an explanation for the mysterious disease. I described the symptoms I could remember to a doctor friend and explained how, crippled, I had to drag myself along the floor. This rheumatologist told me it was probably rheumatic fever, for my left leg was drawn up tight against a frail body—the same leg that had been broken by a fall with my mother when I was barely a year old. Such is fate.

As a student in college, I was probably the youngest in a freshman class of some 170 students. Many of them had to stay out a year or two after high school because of the Depression. I was fortunate to be able to go to college right out of high school. The in-betweeness seemed to be lessening as a factor in my make up. I could hold my own even if I wasn't the brightest student, even if I had to study harder than many others. Almost a quarter of my class had better grades in college. However, more than three quarters had grade point averages lower than mine, yet even that was in-between. One of the greatest discoveries a young man can make is to realize that there are always, in every field and every interest, people who are smarter, stronger and faster than oneself. Should that make one lie down and cry foul? Of course not! To get to do what one wants to do takes a certain amount of grit of which mine was about average! Could fear of poverty—seen among many of my friends' families and in the first cities during the Depression—spur one to achieve financial independence? One must presume so. Once a boy realizes his own limitations, there can be no further obstacles hindering his ability to find a niche, enlarge it, and turn that niche into good fortune—fortune beyond even the wildest of dreams. But the dreams must be there—and to dream one must observe, be adaptable and stay healthy. In-between indeed!

As a student at Wooster in 1936

As a baseball player

In my football uniform

Almost Never First

To paraphrase Andy Warhol, "Everybody has a right to be famous for at least fifteen minutes." This has to do with activities in my own life, yet it has nothing to do with fame; it has everything to do with situations in which I've found myself in some kind of competition with peers. No matter how capable a person is, there's always someone better. That others present themselves as more capable should be a foregone conclusion which strengthens the will to do the very best one can. Gen. Hap Arnold, that Early Bird who was Chief of the Army Air Corps in World War II, once said, "What's hard we do right away. What's impossible takes a little longer." Challenged by time and problems, there is always another chance. The concept that opportunity knocks on the door but once is erroneous; one simply must find more doors, or even erect more doors for opportunity to knock upon. There's an old Pennsylvania Dutch saying: "It ain't what you have. It's what you do with what you've got." That makes plain the idea that other people may be smarter, but if you hang in there you can make some kind of mark. Who could have said it better than Marcus Tullius Cicero: "If you aspire to the highest place, it is no disgrace to stop at the second, or even the third."

The opportunities to excel in one lifetime are many. But there may be few chances that allow one to be first. No one knows this better than I do, so let's look at some. In grade school, as I reached the eighth grade I was studying arithmetic with a teacher who, I much later discovered, was a first cousin of my mother's. This mature man took us on field trips into fields, the woods, along streams, and got our class interested in science. He insisted we pass a comprehensive test on everything he thought we should have learned—a unique idea for schools today but a normal requirement before we could go to high school. I passed about third in a class of eight but failed arithmetic! How in the world could one fail arithmetic? I knew the multiplication tables, how to add, subtract, divide, multiply, and do fractions and square roots. He flunked me on timed tests—no computers then, no slide rules available, only paper, pencil, and an hour to do two hundred problems! It was my first failure, so he drilled me against the clock, and in another week, I passed my first rigorous test in school. The ignominy of that failure didn't faze me at all; it only caused me to get angry at my daydreaming!

In high school, basketball was the major sport, for we had a new school building with a splendid gymnasium and basketball court. A senior generally was chosen captain of the basketball team, but the team voted my cousin, a big rangy junior, captain of the team. Angered at this briefly, I became the high point scorer my senior year—only the following year, that same cousin broke my school record for points scored by many points. He was my best friend, however, so it didn't matter.

In my junior year in high school, the principals of four counties got together for a conference. We knew they were cooking up some kind of mischief as administrators are wont to do. Each of some 20 schools could choose at least five students, preferably seniors, to take an assembled all-day test in Killbuck, Ohio. We met in several classrooms to take timed tests on all high school subjects, yet I hadn't taken my senior classes yet. Out of the 100 students, I ranked fifth in the four county area; almost never first! As a senior, total rewards were compiled, and out of a class of seven, I ranked third, two percentage points behind two girls who were named valedictorian and salutatorian. But guess who was voted president of the class earlier in the year—I was, much to my surprise.

High School
1931
I am on the left

In college, I was awarded my freshman numerals, 1936, in basketball and in baseball—the shortest boy on both teams. Becaue everyone was so much taller and better at basketball, I never tried for the varsity team. But in baseball, I played on the varsity squad as pitcher and center fielder during my sophomore and junior years. I was never first in any sport in college for many reasons—mostly size.

College classes were difficult with challenging professors who helped me grow up to demands I had never experienced. I kept a B+ average—top 25 percent of our class but not the top 20 percent who were recognized with special honors. However, with a special project, I was awarded

Departmental Honors in geology at the College of Wooster. Passing the comprehensive oral and written examinations in geology was easy as my grades were straight A's. In the day-long written comprehensive examination, I scored only 85 percent, clearly in second place among seniors that year. There's always someone better.

As a roofer for eight summers between the ages of 11 and 21, guess who was never boss of a roofing crew? Who was always nailer of the top row of sheets, the lowest ranked roofer of the four to six older men on the crew? It is in working with a gang of men that one learns about pecking order.

Wooster Airport
July 1936

When I learned to fly airplanes at Wooster Airport with Roy Poorman as instructor in 1934-35, I was told I was an above-average pilot, but in spot landing and spot bombing contests, I was the pilot who always came in third to last. In 1947 I failed the flying part of the private pilot's test but passed it later. Earlier, as a model airplane designer and builder, I flew in the American Model Airplane League national contests and in the Wakefield International competitions. I was awarded two major honorable mentions, holding a position within the top 10 percent of all contestants in 1931 at Wright Field, Dayton, Ohio. But at the Mansfield Air races that same year, I won first place for longest model plane flights—a lonely first in many contests.

For many graduate geology students at Harvard University during the depression years 1936-1940, a crowning blow surfaced in the spring of 1938 when the Division of Geological Sciences instituted a new unannounced, two-day, 12-hour, total field, written examination in geology. Despite the Depression, or perhaps because of it, when jobs in geology were virtually non-existent, over 60 students, seniors included, were required to take it and pass with grades of 85 percent or higher. John Lyons, a senior, and I, a second year graduate student at the time, passed this rigorous test and tied for fourth place. Shortly thereafter, at least 15 graduate students were asked to leave Harvard having flunked that same examination.

The U.S. Geological Survey examination for the geologist position in the government was held in the fall of 1939. I scored 76 percent, barely two percent above failure. Sixteen of us at Harvard were urged to take this important examination as practice. Of the 16 from Harvard who took it, 15 passed and one failed; of the 16 from M.I.T. who took this exam at the State House in Boston, 15 failed and one passed. Later it occurred to me why such disparate records occurred; I found out that Harvard professors and Harvard graduates working for the Geological Survey had written this federal examination! Despite the grade of 76, I was offered four different federal jobs in different divisions of the Geological Survey.

My record of almost never coming in first was shattered when I passed the comprehensive oral examinations in geology in February 1940. Many failed the acid test each year and, if they failed a second time, were advised to leave Harvard.

Challenges beset one in any field, wherever one goes. In 1957 I was elected to be national editor for the Bulletin of the American Association of Petroleum Geologists—not much of a feat as I had no competitor during the election. Earlier however, I won the President's award for the best paper published in the AAPG Bulletin by an author less than 35 years of age during 1948—a rare first among many competitors. In 1969 I was asked to run for the new position of President-elect of AAPG under the vastly revised constitution of this largest geological society in the world. No one was allowed to electioneer in any way. My competitor was a well-known author and senior geologist of a major oil company. Would wonders never cease. I won the national election, and to this day, I don't really know why. Probably an uninformed electorate!

In consulting research work and exploring for oil, I won't mention the number of dry holes on exploration prospects I analyzed; nor will I burden the reader with my several successes. But it is noteworthy that I was passed over one year in rising in rank from Associate to Full Professor; also as an active Naval Reservist, I was passed over one year in rising from the rank of Commander to Captain—almost never during the first opportnity did I accomplish these things, but eventually the prize was always won.

In all of this tedious litany of seldom being first, there lies some kind of design. If I ever discover what it is, I may write a book.

Forks in the Road

Not a soul alive has ever avoided forks in the road of life. Far-reaching decisions must be made many times which change the direction or determine a new direction for an active life to follow. Because no man can appraise the effects of these decisions by other people, the reader will recognize that what I write here about such important decisions must be, necessarily, based on my own experiences. Here we go!

To begin, the first fork in the road was none of my doing—what sex I was to be was wholly dependent on my father. As the second child of a family whose first child was a boy, my mother wanted a girl. But born a boy, the best she could do was hope I would someday marry a girl who would be like a daughter. This I did at another fork in the road; I married a girl who had no brother, so my marriage resulted in a son for my mother-in-law—namely me.

There were no forks in the road concerning grade school and high school. I went to a class "C" public school in the village of Berlin. I could have gone to South Bunker Hill School, six grades in one room out in the country where my mother went to school the same distance away. I had no choice in the matter as my father wanted me to be a student with a dedicated teacher, Miss Engel, at Berlin School. So off to school I walked, unable to speak English, for our language at home was a dialect of German Swiss.

The first real fork in the road I encountered was which college and where? One of my mother's cousins had gone to the College of Wooster. It was a liberal arts school,

Presbyterian, and nearby, whereas the top Mennonite school, Goshen College, was far away in Indiana. Anyway, Goshen was at the time touted as a strictly church-dominated school dedicated to educating teachers and preachers, not medical doctors. My father insisted I become a doctor, but my mother suggested I be a preacher; my father won out, so I went to Wooster. It was my decision, for Wooster was well thought of as preparatory for a career in science.

There was an earlier minor fork to consider, but that decision was made by the principal at our new four-year high school in 1929, following my freshman year in the old wooden school building. Mr. Smith, a newly hired fireball of an administrator, was an astute advisor who went over the records of our freshman year and said to me, "I know you want to go to college" (which was true), "so you'll take the academic courses in high school, rather than such stuff as manual training, shop, business arithmetic, etc." At that time I had survived my freshman year with a particularly unscintillating record because I wanted to be a pilot. I spent my time learning all I could about aerodynamics on my own and building model airplanes when I wasn't working as a roofer during the summers. Nonetheless, academic courses were to be my lot: every science through physics, all the mathematics taught at the school, Latin until it came out of my ears, grammar and literature, civics—you name it—plus baseball, basketball and gymnastics.

Getting into the College of Wooster was easy; the college needed students during the Depression and my high school records were within two percentage points of the valedictorian's, and I was president of our senior class. At college I thought of archaeology (which wasn't on the curriculum), pre-med, then coaching, and perhaps high school teaching, which required one to be a coach as well as a three-letter athlete in high school. Coaching looked good until

I talked to old Dr. Fackler who informed me that I'd be required to take at least 21 hours of courses in education—a complete waste of time, for the content was as dull as the three professors who taught the courses. What to do? By that time, in the middle of my sophomore year, I discovered geology, the first real fork in the road career-wise. So began a busy schedule of catch up: two summer school sessions, a tough year as a junior taking biology, and two of the toughest courses in geology. I did this all while working two part-time jobs and learning how to fly airplanes, which I knew would be necessary to become an exploration geologist!

The next fork in the road, more critical than any before, was which university to choose. So I took the U.S. Naval Aviation physical examinations thinking to join the Navy in 1936 after graduation. Then, always wanting to go West, I matriculated at Stanford University for $10.00. My father had laid down the law: "No Navy. You're a Mennonite and we don't believe in fighting." I countered his dictum by saying, "But you were drafted to go into the Army on November 12, 1918, and you didn't plead conscientious objector. Why can't I join the Navy and then go to Stanford?"

Earlier, during my sophomore year at Wooster, my father was told by a salesman friend "Geology is a dead-end. Don't let him study geology!" My retort was "If I can't study geology, I'll come home, sit around and do nothing." A whole other story involves how he relented even though he was a domineering man. When I threatened to join the Navy and then matriculate at Stanford in May of 1936, he said, "Stanford is far away.Your Model A Ford might not get you there. Isn't there a superior school of geology closer by?" I said, "Yes, Harvard" (which was then the best in North America, perhaps in the world). So he said, "Why don't you go to Harvard? I'll help you borrow the money at the bank!" I had won; the threat of U.S. Navy duty and going to a school far away had done the trick. Forks in the road! I went to Harvard.

The first half year I found Harvard to be an unfriendly school despite my friendship with the professors and a physical education instructor named David Hyde who became a close friend. I left Harvard to try a side fork: go to University of Pittsburgh, get a masters degree and go to work for an oil company. The schedule at Pitt was impossible at half year. Dr. Richard Sherrill didn't want me to take extra courses at Carnegie Tech, so I gave up that fork in disgust. With $105.00 to my name and a Model A Ford, I came back home and took off for Oklahoma in February of 1937.

I lived in Tulsa at a rooming house and tried calling some oil companies. All turned me down—another story, another fork in the road. One company, Shell, said to "Come in for an interview," and I did. But it was a dead end—no job as geologist or even as geological draftsman. Ward Bean, the interviewer, said, "Wait. Go see our geophysicist, Dr. Gerald Lambert. We're starting a new seismograph crew." I got the job at $90.00 a month; I was down to my last $20.00—barely enough to go back home if I couldn't get a job. I was saved by the bell and Shell's geophysical department.

Six months later, with wages at $105 a month, I realized being a jug-hustler on a seismic crew could only lead to being a crew chief—for me a dead end as a geologist. I found a second fork in the road and went back to Harvard. I wrote to Harvard asking "May I come back?" The chairman of the Division of Geological Sciences at the time, Dr. Kirtley F. Mather, whom I had met briefly only once during my first half year there, fired a letter back to me saying "Yes, come back to the university!" They must have been hard up for students in the Graduate School of Arts and Sciences.

Later, a letter from Shell arrived: “We have a job for you as a geologist—$135.00 a month.” By this time, Dr. Mather was my advisor, for he was the only ex-petroleum geologist on the staff. He told me in no uncertain terms, “Don’t you dare take that job! You haven’t even taken your final all-field comprehensive examinations yet; if you leave you’ll never come back.” Another side-fork was avoided. My first full year, 1937-38, I studied full-time with no outside jobs, living on borrowed money. I stayed at Harvard between 1938 and 1940 as an Austin Teaching Fellow, slinging hash at a rat house, skimping along earning my total way, still in debt to the bank back home for my first year and a half at the school.

Bicycling at Harvard
September 1939

Forks in the road were coming along fast, and it was difficult to choose between them. In 1938 I decided that I wanted to do a dissertation on an oil field stratigraphic project and not map river terraces for Dr. Kirk Bryan, nor map ring dikes in New Hampshire for Dr. Marland Billings, nor map a mine for Dr. Donald McLaughlin, who was my first advisor in the fall of 1936. By February 1940, I passed the comprehensive examinations, got engaged to Dr. Mather’s daughter, Florence, and decisions had to be made. I wrote to many oil companies. No dice because no one was hiring geologists. Along came a letter from Shell explaining that a Mid-Continent vice president was coming to Harvard to interview me with my dossier in his hand! That fork in the road was easy. I had also been offered four different positions with the U.S. Geological Survey because I had passed the civil service examination in Boston, and an offer to teach at Syracuse University came along, which I turned down. I interviewed with Vice President Louis Roark of Shell Oil Company and, at last, two weeks later, I had a position as geologic trainee in a new program established by Roscoe Shutt, the Mid-Continent exploration manager for Shell Oil Company. The salary was $160 per month!

The decision to marry Florence and take her along to Oklahoma was easy. I was on the way. Pearl Harbor threw us into World War II. Incensed that Japan had the nerve to shell a refinery at Long Beach and having already decided to start a family, war or no war, I took the Army Air Force examinations to become an attack glider pilot, as I had lots of light plane flying experience. Application to the U.S Navy as an aircraft facilities intelligence officer won out in June 1942 after a U.S. Navy officer in March had insisted that I become a radar officer. The fork in the track to the sea was clearly the Bureau of Aeronautics as an AV(s) officer in the hydrographic office of the Navy rather than as a glider pilot in the newly developing Army Glider Corps.

Shell had reserved my job as research geologist during my Naval duty in World War II, duty which took me from Washington, D.C., to the Canadian Arctic, the Western Pacific, and Alaska, and no decision was necessary. Had there been no return possible to Shell, I would have moved my family, now with two children, to Alaska, an inviting frontier.

At work on my doctoral thesis
in the Arbuckle Mountains, Oklahoma
1941

In 1947, after final completion of my Ph.D. dissertation—written during all my spare time while on regular exploration duties with Shell in the Mid-Continent—and following the receipt of my degree from Harvard, I was offered teaching positions at both the University of Nevada and the University of New Mexico. Because of a love for the Southwest and Mexico and the guarantee of housing, I took the teaching position as assistant professor at New Mexico—the wise choice in what was a very important fork in the road. I could do consulting work in petroleum exploration in the truly frontier Four Corners region, the area in which Arizona, New Mexico, Utah, and Colorado meet.

Then came many minor forks involving academics. I turned down all other offers to teach or be an administrator at five other universities—not one with a climate as salubrious as that of New Mexico (an ideal place to raise a family). I stayed, despite several bitter academic experiences—another story. Elevated in rank to full professor with heavy teaching, publishing and research responsibilities, the many decisions to stay were a straight track with no forks until the time when my wife and I decided to retire early and leave the academic maelstrom. Then came the easiest fork: to decide whether to stay in New Mexico, move to Hawaii or, in fact, any other place. No problem—I chose to stay in New Mexico, my geologic home.

One of the barracks—
Our first home in Albuquerque—
Varsity Village

By mutual decision, Florence and I have travelled throughout the world. As we look back, we both wonder: what if I'd become a Naval aviator in 1936 (probably shot down in the South Pacific); what if, earlier, I'd taken the fork leading to Stanford (I wouldn't have met Florence); what if I had taken the job with Shell in the spring of 1939 (no marriage to Florence); what if after graduation from high school, I had decided to follow my mother's wishes and go to Goshen College to become a Mennonite preacher (heaven forbid!); or, even earlier, what if I had decided to remain a roofer in my father's successful company (mental suicide!)?

Every active person makes major and many minor decisions that fashion a life. Many make unhappy decisions. Most of mine have been fortunate, and I, for one, would not change my life one whit—or wish to live it over again any differently. In fact, it's been a heck of a life. How lucky can one person be?

Several Careers

It is not common for many men to have several interesting careers. Most go to school, graduate at whatever level, go to work, marry early, have kids, struggle through a life of frustration, retire at 65, and die before 70, having never done all the things they wanted to do. As they "shuffled off this mortal coil," unheralded, unheeded, and soon forgotten, they may nevertheless have left a legacy of rectitude and greatness in their children. Some marry several times, in itself multiple careers. Others never marry, die in apparent spiritual poverty after a non-auspicious life of little achievement, yet they leave bundles of money stashed in every available hiding place in their humble living quarters. One reads about such men and women after they are gone, and it becomes obvious that they have had several fascinating careers. In a few cases, they leave reams of writing. If they die alone but have left ungrateful children, such writings may become historical documents of the human condition of a lost time.

Permit me to write a bit about my own careers. As a boy aged 10, out of the fifth grade in a small town in Ohio, my father told me one mid April day, "Now you must go to work with my roofing crews this summer." I averred that I didn't want to, but I didn't prevail. I went along for a week or so. For a sickly boy, the Ohio weather was too cold and wet that April of 1925 so I caught bronchial pneumonia. This sickness, almost the end of my rather sickly childhood, forced my mother to keep me at home during the summers until I was out of the sixth grade. My career as a roofer lasted from that 11th summer until I was past the age of 21, nine years, having missed eight weeks of the summers of 1934, for college summer school, and 1935, for geologic field school in Virginia.

Summer field camp in Virginia
1935
I am on the left

A second career began in the fall of 1934 when I began to learn to fly at the Wooster Airport with Roy Poorman, a tough Germanic flight instructor. Earning only 30 cents per hour through part-time jobs at the College of Wooster but paying $6.00 an hour for the lessons, I soloed a Waco ten, Ox5 powered biplane on June 10, 1935. This began a flying career of some 40 years.

I delved into several other fields, many already mentioned: my career as student when I was educated as a geologist at the College of Wooster and at Harvard; the brief time I spent as a jug hustler and surveyor on a seismograph crew; the period in which I worked for Shell Oil Company as a research geologist; my career in the Naval Reserve, extending until 1975 upon retirement as a Captain USNR; and those years from 1947 to 1976 as Professor of Geology and in 1981 as Professor Emeritus of Geology at the University of New Mexico.

As a Wooster graduate
1936

I had two more careers to go—and perhaps a third or even a fourth! Between 1947 and 1976, I also spent time as an oil and gas research geologist on consulting work for many major and minor oil companies to finance a career in publishing articles on geology in local, regional and national periodicals. For five years, from 1970 to 1975, I served the American Association of Petroleum Geologists as their first official president-elect, president, chairman of the Advisory Committee, and chairman of the Officer Selection and Honors and Awards subcommittees (a career in itself). Then began a continued career as a naval historian of activities in World War II. While in active retirement, careers of travel commenced in 1976. The fourth? Maybe I will be able to continue a writing career. Who knows?

Woodward, Oklahoma

Out Where the West Begins

Who knows which fateful moments or deep experiences may color our lives? As a boy of eight in Ohio, having learned to read English only three years before, I discovered books. One of the first was the story of a young man who got into a fight in a bar in Illinois. He killed a man, or thought he did, and escaped to St. Louis, then the Gateway to the West. Fearing capture, he moved Southwest into the deserts and wandered about in the mountains. His experiences affected my whole life, and the desire to go West was overwhelming.

Camping on our western trip
1937

In early 1937, years later, I left graduate school after a half year in the East and went to Oklahoma to work on a Shell Oil Company seismic crew. Forays westward to New Mexico and Colorado on three-day weekends primed the pump for a grand circuit in late August in my Model A Ford, camping with a college buddy to save money. We traveled through Denver, Salt Lake City, Reno, San Francisco, Los Angeles, San Diego, and Phoenix—the West had me in its grip. The next year I spent a summer in Idaho mapping the geology of a silver mine at Ramshorn Mountain. This was followed by two winters back at graduate school in the sodden, snowy East. It took 10 years, one marriage, two babies, a four-year war, and work for an oil company two years before and two years after the war, before I could live permanently in the real honest-to-goodness West.

But where does the West begin? Lewis Wright, the cowboy who was my mentor on herding cattle when I worked for him as a weekend cowhand, maintained that the Big Thicket of east-central Texas was "where the pavement ends and the West begins." Through many trips across the country by car and motorhome, by airplane and train, I've tried to define where that line between East and West really exists. Is it a line or a wide zone? So I started marking up maps during these transcontinental trips, looking for and recording all the things which I thought were typical of the West. Earlier I realized, from photographs and maps, that one could use criteria that yield sharp boundaries. The 20 inch rainfall line, well defined by average rainfall over a hundred year period, gave one measure, for to the west it was dry and to the east it was moist. Or one could delineate the eastern edge of the High Plains province. If a lazy person wants to choose well-surveyed lines, he should use the 100th meridian or the boundary between the Rocky Mountain and Central Time Zones. If you live in the East and don't really care about the West, choose the Mississippi River; it would be a logical boundary (so most people think). Of course, New Englanders believe the West begins at the Hudson River!

The West

After 35 years of living in the West and looking for what may define the West, I've come up with all kinds of stuff I think makes the West the unique place that it is. What really started me off on this search was a poem I found on a postcard in 1937. That poem, long a part of our eminent domain, free of copyright, was written by Arthur Chapman and originally copyrighted sometime before 1936 by J.R. Willis. This poem, simple in its cadence, sincere in its meanings, has become a classic. Here it is:

"Out Where the West Begins"

Out where the hand-clasp's a little stronger,
Out where the smile dwells a little longer,
That's where the West begins.

Out where the sun shines a little brighter,
Where the snows that fall are a trifle whiter,
Where the bonds of home are a wee bit tighter,
That's where the West begins.

Out where the skies are a trifle bluer,
Where friendship ties are a little truer,
That's where the West begins.

Out where a fresher breeze is blowing,
Where there's laughter in every streamlet flowing,
Where there's more of reaping and less of sowing,
That's where the West begins.

Out where the world is still in the making,
Where fewer hearts with despair are breaking,
That's where the West begins.

Where there's more of singing and less of sighing,
Where there's more of giving and less of buying,
And a man makes friends without half trying,
That's where the West begins.

In 1938, driving across Wyoming, there were small black balls rolling down the highway. What's this? We stopped to see a sight I'd never seen in the East: dung beetles, also known as tumble bugs. A large one with long hind legs and a small one with large hind legs were rolling a ball of cow manure down hill on the asphalt pavement. The graduate students with me suggested we make a research project out of this phenomenon and trace these balls back up the hill. Sure enough, we came to a cow pile on the road where dozens of dung beetles were busily separating cow manure into small patches, mixing them with some sand and dust and kneading them into balls, using all their legs. But why were the big beetle and the small beetle working together? It became obvious that these were a large female and a small male. Then we saw that they were pushing the ball with their hind legs, and if the road was steep enough, the ball would start to roll.

The small beetle, obviously a lazy but intelligent fellow, would hop onto the ball and ride it until it stopped—with the female running alongside, too smart to get bumped around having the manure ball roll over her, as it did over him. One could almost hear her laughing as she ran alongside as her beetle mate hung onto the rolling ball for dear life. Thence, fascinated by this truly Western phenomenon, we traced the caravan of rolling balls of dung to where they stopped in weeds beside the road. Having eaten her fill, the female began to lay eggs in the dung ball which was secreted under a hollowed out rock, thus providing food for newly born dung beetles hatching out of the eggs. How did they learn to do this? Not being entomologists, we puzzled over this. Later we found that this also takes place in West Virginia—but at least that state is west of Virginia!

To continue this analysis of where the West begins involved recognition of the sharp angles of mesas and buttes, their flat tops and steep sides of sandstone, with shale making a gentler slope. Topography is the first thing one should notice. Dry arroyos abound in the West, running bank full during the heavy desert rainstorms that march across the parched landscape. I once saw a car tumbling down one of those dry arroyos in a wall of water causing picnickers to flee for their lives as the wild flow carried their car away. The rain fell heavily miles to the south. The great, but often dry, rivers draining the Rocky Mountains eastward across the Great Plains onto the Prairie Plains province provide millions of tons of sand which blows out of the river bottoms in dry spells to create vast bands of dunes. Many of these areas of sand, downwind from rivers such as the Cimarron, the Canadian and the Raging Red, were formed in glacial times of abundant melt water and a wetter climate. Now stabilized by piñon trees and sage brush at elevations over 7,000 feet, as well as juniper, yucca and grass at lower elevations, these vast sheets of sand suffer blow outs to create modern sand dunes. In extreme drought, sand and fine silt blow eastward in great sandstorms that roll across the country like mighty steam rollers. Such is the West—a windy place! Many creeks are named Cottonwood, Dry Gulch, Brushy, Muddy, and in one place in Utah, there's an Evacuation Creek, so named by sensitive geographers in Washington, quite unlike the name "Shitcreek" given that creek by early settlers. In the West, steep talus cones of fresh rock extend off steep cliffs. They are fringed in the interior valleys by alluvial cones and fans which, in places, have their edges flooded by playa lakes that dry up to become salt pans. No doubt about being in the West when the four to eight inch rainfall cannot support streams to cut through the rims of downfaulted basins. On bare surfaces, tinajas, hollows filled with rain water, have saved the lives of many animals and desert travelers.

Wind-blown sands
Oklahoma
1937

One can't help but be amazed at the wide spaces of open range, so it comes as a shock to see barbed wire fences with gates made of poles strung with what cowboys call bobwire. As irrigation takes over more and more prairies, barbed wire is replaced by woven fence, pole gates by aluminum gates, some of which are designed to be opened by pulling a release without getting off a horse or out of a pickup truck. The Old West Longhorn has long ago been replaced by Herefords, Ayrshires, Holsteins, and Angus fed in great smelly feed lots out of mixing towers filled with artificial fatteners. As one crosses the western side of the East, coyotes, jackrabbits, badgers and prairie dog towns enliven the scene. Rattlesnakes are underfoot, cowboys in chaps ride paint ponies and wave their hats, and barns disappear to be replaced by pole loaders that gin hay up on great stacks. Here, too, changes are rampant. Large machines cut hay in irrigated or bottomland fields along river courses, roll it up into giant rolls, or pack it into great blocks stacked along fences. Some are wrapped in plastic by automatic machines and stew in their own juice to become ensilage—a new development in America, but long known in Merrie Olde England. Alfalfa, corn, soy bean, and sugar beet fields are appearing at an alarming density far west of where the West begins, wiping out virgin prairie and sage brush plains so typical of the Old West.

The vegetation of the Old West involves bunch grass, chamisa, fields of cactus, rabbit brush, Mormon tea, that pernicious weed the juniper and, when altitude plus accompanying rain allows, piñon, spruce, and other evergreens. So as you drive the blue and red roads, staying off the freeways, you see houses of adobe or rock with flat roofs, or cowboys on horses and in pickups with empty beer cans rattling around in the back. You also witness Indians, in wagons in the old days, now riding in bright muddy, dusty pickups, usually driven by women in the matriarchal society where the chiefs are still men. Large cattle trucks also hurtle down the roads. The trail drive is a rarity except in low prairie to high mountain meadows in the spring and back down again as winter nears. Bandanas, cowboy hats, lariats on saddle horns, Winchester rifles in saddle scabbards or across the back windows of pickups, and cowboy boots on tanned men and women are common. Calves being roped to be castrated, the result being delicious mountain oysters fried on the gas-fired branding iron heaters, rodeos in small towns and large, the great roundups with a cowboy trail cook and his chuck wagon, and flap jacks, bacon and biscuits mixed by hand in the top of a flour sack are signs of the West.

Now you may say, some of these signs of the West exist in the East, such as cactus on sand dunes on Cape Cod, drug store cowboys on big ranches in Florida, vast fields of alfalfa, and tumble bugs in West Virginia. Perhaps there are others. But when you find the things listed above together, combined with distant vistas of rolling plains and stark blue mountains with clouds boiling up over them into thunder heads against vast blue skies, YOU KNOW YOU'RE IN THE WEST!

Part Two

BACKGROUNDS

Errant Peasants
A Pennsylvania Dutch Code
Grandparents
Aleck and Catherine
Steffy and Betsey
Parents
Al and Lizzie
Kirtley and Marie
Dad as a Youth
Travels with Father
Foiling the Pickpocket
Dad's Friends
A Mouse in the Soup
Vindiction and the Human Spirit
The Bell
Crack of the Bat
The Aviators

Errant Peasants

Fortunately, all of us have ancestors—some fewer than others because of intermarriage of family lines far back in history. This was particularly true among families who were close knit, lived in language and topographic enclaves, or did similar work, such as farming, near each other. Some ancestors were better than others in morality, work habits, religious observance, and family record-keeping. Family Bibles among Mennonite and Amish religious groups took the place of parish records and were where official records of births and deaths were recorded.

According to the state archives in Berne, Switzerland, the Wenger family was a family of means early in its history. The name can be traced back to the 13th century, long before the Anabaptist movement of the 16th century. Belonging to a discrete set of families of several origins in Germanic Switzerland, the name appears prominently in state records only as early as the 14th century. Many Wengers were burgermeisters, land politicians, tradesmen, and the family was prominent enough to have crests. In the Wertheim area, they became enemies of the Catholics of Wurtzburg. When taxing systems were inflicted on Wengers by the dominant Catholics (the royalty and largest land owners), they began to flee from the agricultural plateau of the Bernese Oberland southward into the Alps. As these families grew, the Catholics followed them into the foothills of the Alps to press them into enforced labor, charging fees for grazing and hunting and forcing many into unwanted military service. They had to pay water, pannage, plowing, death, and food taxes; later, they were to pay hay-crop and handicrafts taxes, and taxes on horses, cows, pigs, and chickens. Many could be paid in lieu of enforced labor, such as a work tax, a chimney tax, hearth (property) tax, tithes, ground rents, special assessments, and even one prince's onerous personal tax! All these infringements on freedom, dearly held by the early Catholic Wengers, drove them to become seekers of religious freedom by joining the revolutionary Anabaptist movement. Unwilling to fight, these sturdy people became vehemently anti-Catholic. Seeking to escape Switzerland, they unwillingly paid emigration taxes to the Swiss Catholics and began moving down into the Schwartzwald and Alsace-Lorraine areas of pre-Germany and France. They remained there for a century before beginning their move to America in 1717. Major migrations of the older Wengers took place between 1717 and 1840. Among them were my forebears.

To this day, the Mennonites, followers of Menno Simons, and the Amish-Mennonites, followers of Jacob Amman, believe in avoiding the government, not serving in armed forces, holding the belief that they have a right to read the Bible rather than have it inflicted on them by priests, and upholding a complete separation of church and state.

Into this milieu of the early 1700s, there was born a man named Johannes Wenger on a farm on the Wengernalp, a small Alp with broad grassy meadows on the side of the Lauterbrunnen Valley. Earlier, several of the Wenger families, many of whom gathered on these high mountain grazing areas of Switzerland, established the village of Wengen. When this home environment became untenable because of political and economic problems, Johannes moved out of Wengen, the last Wenger to do so. We know little of his family—how large it was or exactly when they left. One son, Jacob, and his brother, Christian, were probably born either in Wengen before the family left or in the Interlaken area by the Thunersee.

Jacob was my great-great-great grandfather. We believe he lived between 1750 (?) and 1789 in either Berne or the Schwartzwald area. He married and had a family. An older son—born in 1786—was named Christian after his uncle. According to one family legend, Jacob, his brother Christian and Christian's family went by ship to Philadelphia in 1789. They made their way through Lancaster to Cambria and Somerset counties in southern Pennsylvania and settled on land in Brothers Valley. In 1801 Jacob recrossed the Atlantic to his family, who had been living with relatives, and started with them for America. While on that voyage with his wife and children, Jacob died, and his body was buried at sea.

Whenever I cross the Atlantic (by now some 15 times by ship and airplane) to work overseas or on vacation, I think of the rigors these gutty ancestors of mine underwent to gain religious and economic freedom.

Saddened beyond relief, Jacob's widow, two sons, Christian and Joseph, and daughter, Catherine, landed in Philadelphia with little money. They travelled to Lancaster, staying with relatives long established there, and then moved on foot, by packhorse or by wagon with other relatives westward across the Appalachian Mountains to the then wild frontier of Cambria and Somerset counties in the broad Brothers Valley high on the Appalachian Plateau. Expecting to find out they owned the property that Jacob had acquired, my pioneer Mennonite ancestors found that Old Uncle Christian had paid the taxes for his brother Jacob and, by that subterfuge, had taken over the unperfected property. Young Christian—15 years old when he crossed the Atlantic—his brother, Joseph, his sister, Catherine, and a third son, Jacob—born in Pennsylvania in 1802—were left destitute. In order to live and stay together, they were taken in by kindly

Old Order Amish immigrants. From that time in 1802 to 1900 or so, the children's immediate family were Old Order Amish. Many of the descendants of Joseph, Catherine and Jacob are today Old Order Amish. Shortly before the Civil War, Christian and his son, Moses, decided to add a "D" to Wenger, and from that time on, the name has been spelled "Wengerd" in defiance of Old Uncle Christian's perfidity and his maltreatment of the late Jacob's family.

The Miller Amish and Mennonite graveyard southwest of La Grange, Indiana, is the site of Christian's grave; N.E. Section 9, T36N, R9E marks the exact spot. A new white marble tombstone was placed on the grave about 1960. On this tombstone the name is misspelled Winger, an Amish-Mennonite spelling—clearly an error. He was born January 3, 1786, in Switzerland and died June 30, 1882, in Indiana at over 95 years of age. He was an angry man whose nickname was "*Der base Grischt*," (the angry Crist). In the cement at the base of the tombstone is inscribed the phrase "To America in 1801." That Christian was my great-great grandfather—a tough ancestor indeed!

The Wenger migrations as Wengerds from Pennsylvania's Somerset and Cambria counties westward to Ohio are not adequately documented as to dates nor as to who travelled with whom. My father told me a story about his grandfather Moses Christian Wengerd. As a young man, my great-grandfather started walking to Ohio with an acquaintance who had a horse. They would take turns riding the horse and would sleep on the ground at night. Early one morning on about the third day of the journey, my great-grandfather, then a single man, woke up to find the man and his horse gone. So Moses had to walk most of the way to the Sugar Creek Valley in Tuscarawas County in Ohio. Roads consisted of old Indian trails—later after 1790, they were widened by hard hand-labor into wagon trails. In the Sugar

Creek Valley, Moses, a short man, married a tall, beautiful Amish girl named Magdalena Yoder, and together they sired numerous children—among them my grandfather Stephen, of whom I have many memories. It seems logical to believe that these pioneers lived happy lives. This leads me to a quote by Barbara Dickson, a well-known Scottish songstress: "The key to happiness is coming to terms with your background."

At age 2

A Pennsylvania Dutch Code

The Pennsylvania Dutch are Germanic Swiss who emigrated to America in the 18th century from the Bernese Oberland in northern Switzerland and from Bavaria in the 18th and 19th centuries when country boundaries were flexible and there was no united Germany. They had come through Bavaria to the Schwartzwald and into Alsace-Lorraine and the Netherlands seeking freedom among the Huguenots, also in revolt against the Catholics. Earlier, in April of 1598, the Edict of Nantes had been promulgated by the French king, Henry IV, granting religious liberty to his Protestant subjects. The edict was greatly disliked by the Roman Catholic clergy, and because of this dislike and the strong political position thus given to the French Protestants, the edict was revoked by Louis XIV in October 1685. Believers in strict separation of church and state, they developed great antipathy to war and killing and fighting in armies. They believed that they could talk to God and study the Bible without intercessions by priests, and they eschewed infant baptism, believing that each person should reach an age of decision before baptism.

Out of this melange of Anabaptists, there developed a strong individualism. Such individualism helped them spring at the offers William Penn made for them to emigrate to America and settle on land grants in Pennsylvania—long before it became a commonwealth of the British Empire. Between 1723 and 1810, most of their families moved across the Atlantic to Philadelphia amid severe loss of life through shipwrecks, disease and cruelty of ship captains.

Surprisingly, the Old Order Amish who believed the Mennonites had become too modern and too worldly too fast nonetheless live peaceably with them wherever they are found together. They are bound by the common complex German dialect of Pennsylvania Deutsch. Although they may bicker within families, roust each other as families, and develop offshoots of both churches with all kinds of variations of style and beliefs, these people of my background developed a code of living worth noting. They had learned the true value of time, the success of perseverance, the pleasure of working, the worth of character, the value of knowledge, and the observance of God's Word. All are aware of the importance of reputation and are alert to judging the quality of character. The code is simple and taught in every family: "For successful living, tell the truth, work hard, save your money, pay your debts, help others, lead a joyful life, and sleep in your own bed!"

The following maxims, slightly modified, were assembled by Senior Bishop Andrew Hershberger of the Beachy Amish-Mennonite Church of South Carolina. They are Germanic Swiss concepts adhered to by most people known as the "Pennsylvania Dutch." They go as follows:

1) *Keep good company, or none. Never be idle. If your hands cannot be usefully employed, cultivate your mind.*

2) *Always speak the truth. Make few promises. Live up to your engagements. Keep your own secrets—if you have any.*

3) *When you speak to a person...look him in the face. Good company and good conversation are the very sinews of virtue.*

4) *Good character is above all things else. Your character cannot be essentially injured except by your own acts. If one speaks evil of you, live so that none will believe them. Never speak evil of any one...be just, before you are generous. Keep yourself innocent if you would be happy.*

5) *Drink moderately, if at all, of intoxicating liquors; do not use tobacco in any form.*

6) *Ever live (misfortune excepted) within your income. Make little haste to be rich if you wish to prosper. Play seldom at any game of chance.*

7) *When you retire at night, think over what you have done during the day. Small and steady gains give competency with tranquility of mind.*

8) *Avoid temptation; fear you may not withstand it.*

9) *Earn money before you spend it. Never run deeply into debt, unless you see a way to get out again. Never borrow, if you can possibly avoid it. Save when you are young...to spend when you are old.*

10) *Do not marry until you are able to support a wife.*

11) *Believe in God, and keep the Ten Commandments, for this is the whole duty of man and woman.*

Elizabeth (Betsey) Wengerd
with granddaughter Carol Ann
July 1933

Grandparents

This introduction to four people who have had remarkable influence on my life seeks to outline some information on their backgrounds, certain vital data and the surroundings in which they lived long, productive lives.

My mother's parents were Alexander Elias Miller, born in 1852, and Catherine Miller, born in 1859. They were married on January 11, 1880. Father's parents were Stephen Moses Wengerd, born in 1853, and Elizabeth Shetler, born in 1860. They were married on December 7, 1878. Both sets of grandparents were born on farms in Tuscarawas County, Ohio. All were Germanic-Swiss descendants of Mennonites, people who left Switzerland, boarded ships in the Netherlands and came to Philadelphia between 1723 and 1801.

Linguistically, their forebears spoke German because their origin was in Bernese-Oberland and in southwest Germany on the plains and in the foothills of the Alps. Moving northward through the Rhine Valley of Germany to Holland, their language took on words from ancient München, from the Schwartzwald, from Alsace (comprised of some ancient French and German), and tinges of Dutch—in fact, the Alsatian-Germanic dialect is similar to modern Pennsylvania "Dutch." In the past 50 years in America, Pennsylvania German has changed from a dialect to a distinct language with its own grammar and dictionaries put together in the Lancaster region by highly educated Mennonite linguists. Over a million people speak this language in America today.

Pennsylvania Dutch, the language I spoke until I went

to school, is the only language I spoke with my grandparents until they died. I still speak it as a kind of *Niederdeutsch* easily understood in the München area; I mix it at times with the *Hochdeutsch* I learned in college.

Everyone who knew these four grandparents used their nicknames. They all knew each other as children of Amish-Mennonite parents. The Millers were Aleck and Catherine; the Wengerds were Steffy and Betsey. I will limit my description of them to their profound effect on my own life, directly and by example. Their biographies could fill two large volumes. I can still hear their kindly voices and feel their presence daily in my own life.

Alexander E. Miller
November 15, 1852 - October 19, 1925
Catherine Miller
April 22, 1858 - January 10, 1930

Aleck and Catherine

My mother's parents lived in the community of Bunker Hill within a mile of our home near Berlin. It was a large farm where Aleck and his children worked at growing hay, wheat, corn, and oats and overlooked a large orchard and a spacious garden supervised by Catherine. No one ever had a nickname for Catherine, other than "Grandma," although all the children—all ten of them—had familiar nicknames. The family was well-to-do, even considered rich, despite its size. Aleck was a shrewd businessman who did roofing and contracting, travelling by wagons pulled by fleets of horses out of Bunker Hill with roofing crews. Two of the daughters never married: Mandy died young, and Matty had a child out of wedlock. Matty's child was raised as Catherine's daughter and was loved by all. She was prettier than all the rest of the family, except for my mother, Lizzie, of course. All the girls did farm work: pitching hay, shocking wheat, cutting and husking corn, milking cows, and slopping hogs. They helped Grandma cook, clean house and bake those luscious giant loaves of bread in a large wood-fired outdoor oven beside the summer house. Aleck and Grandma had welded this large religious family into an efficient team, just as Aleck had trained his roofing teams to do roofing more effectively than anyone else in the business up to that time. Most of his clients were Amish farmers.

The farm lay north of Lion's Creek, and because it

was near my home, I came to know these stolid people far better than my father's parents. For reasons I never understood, I never saw my two sets of grandparents together, though they had played together as children in the Sugar Creek Valley after the Civil War. The reason was that there was considerable antipathy between Aleck and his son-in-law, Al, my father. Aleck sold his roofing business to my father in 1910. The notes were paid off quickly. My father, as a self-trained engineer, was much more efficient in team-roofing, doing large barns, houses and straw sheds in one day. He enlarged the business to include farmers' supplies, such as feeds and fencing, and services like putting in hay tracks, installing eaves troughs and working on myriads of other things that required techniques beyond the abilities of individual farmers. Seeing this early success by his son-in-law led to jealousy on Aleck's part.

The early birth of my older brother Owen four years before led to considerable discord between Aleck and Al. Allen had married my mother when he was 19 and she 17, in April of 1906, and Owen was born in August—a bit too early by all counts. Aleck then did something which proved to be sly and unethical. In spite, he started another roofing business to compete with my father. It was unsuccessful; the ploy didn't work. Despite friction, family solidarity was of the utmost importance, so my father took in as his partner Alvin Miller, a man who had married another of Aleck's daughters, my mother's sister, Sevilla.

Aleck was in all kinds of businesses. He sold gravel out of the glacial sand and gravel deposits on the farm. He raised cattle and had a dairy herd. He bought large properties and built the large mercantile-grocery store in Bunker Hill. He also brought a cheese-maker over from Switzerland—the Schriers family—so local farmers, and Aleck himself, had an outlet for the milk from their dairy herds. Located on the southeast corner of Aleck's farm, the Bunker Hill cheese factory is today one of the largest makers of Swiss cheese in Ohio.

Aleck was a rather dour character. I knew him from 1917 until his death on the 19th of October in 1925. He had white hair and beard; he was about five feet seven inches tall, solidly built, and was a worker who knew no fatigue. He became one of the richest men in Holmes County through all of his business activities. I am convinced that my father owed much to Aleck, for he was an example of the successful entrepreneur—Father didn't have such an example from his own parents, Steffy and Betsey. Not only that, but when the chips were down, he married the boss's daughter, my mother, Lizzie.

Catherine was a saint. She was about five feet three inches tall, heavy, and a woman of amazing kindness. She loved all of her many grandchildren intensely; she always had candy for them in the deep pockets of her overskirt which covered her typical somber Amish-style dresses. Grandma was a Mennonite of great religious conviction. She was Aleck's second cousin. Aleck was an Amishman; Catherine averred that she would marry only an Amishman because, she said, "They're hard workers and smart. But right after I marry such a man, I'll make him join the Mennonite Church!" And she did just that. Other than my many personal visits to see Aleck and Catherine and my aunts and uncles, there were many occasions when I really learned to know my mother's parents. Almost every Sunday after church, this whole family would gather at their farm to visit, play games and eat dinner together. This tradition went on from 1917 until the death of Grandma Catherine on the 10th of January, 1930.

The Amish Church could not hold the *meidung* (shunning) against Aleck, for he provided too many services to Amish farmers, and too many of them were related to him and to Catherine. Shunning was a medieval practice handed down from the early days of the Mennonite Church, even to this century. When I wanted to join the church to which Florence belonged in Newton Centre, Massachusetts, I wrote to the minister of my church in Berlin asking that he send to Florence's minister a statement confirming that I had been baptized in the Mennonite Church, was a member in good standing and had his permission to join the church in Newton Centre. No such statement arrived, but Florence's minister decided to accept me anyway, and we were married. Years later when I was visiting my boyhood home, I was able to gain access to the membership rolls of the Berlin Mennonite Church. My name was still there, but a line had been drawn through it! It was this practice in our Amish and Mennonite Swiss community which prompted both sets of my grandparents to leave the Amish church and become dedicated Mennonites.

My greatest acquaintanceship with my Miller grandparents came much earlier, however, when my mother was pregnant with my brother Wilmer. The August before his expected birth in September of 1919, I was taken to Grandpa and Grandma's house to live for three weeks. I didn't know why. No one told me about the new baby coming. To forestall all the difficult questions they knew I would ask at the age of four years, it was deemed best that I not be at home.

I had only a touch of homesickness. After all, these were the grandparents I had known since I was two years old. Besides, Grandma's daughter, Katie Ann, who was 28 years old and unmarried, took care of me. I slept in her bed; she romped with me and saw that I washed my face each morning. I helped Grandma bake bread. I watched as she placed the large oak paddle with 12 lumps of bread dough into the great oven which I helped stoke with firewood put into the burn box. We then watched those golden loaves appear as she drew the baking paddle out of the oven. Ah, that fresh bread—chunks torn off and soaked with the rich butter Grandma churned out of the cream I helped to separate by turning the centrifugal separator handle. All this was pure heaven for a four-year-old. I had no idea I was to have a chubby little brother. I didn't miss my bossy older brother, Owen, one bit. I was spoiled rotten by my live-at-home aunts—Mattie, Mandy, and Katie Ann—not to mention a doting grandma and my normally serious Grandpa Aleck. Every evening after supper, I'd sit on his lap at the supper table, and we'd all visit as my aunts would "redd up"—wash and dry dishes, that is. The supper table was in the large harvest kitchen which opened at one end to the plain but comfortable living room. Years later, my Aunt Katie Ann told me that I was Grandpa Aleck's favorite grandson. He joked with me, told me stories, bounced me around, laughed a lot with me, and made me totally forget my own parents at our home only a mile away!

One day during those three weeks, he said, "I'm driving to Saltillo with old Bert hitched to the buggy. Do you want to go along?" We started out early. Old Bert, retired from heavy farm work, was now used only to go visiting or to haul milk to the cheese factory. She was a member of the family. We plodded through Bunker Hill, Berlin, down toward Charm, and out along the dusty roads on the wooded low Appalachian Plateau ridges to Saltillo in south-central Holmes County. It took three hours to go the twelve miles. Aleck was to see Ike Hummel about some cattle Ike wanted to buy for his slaughter house in Saltillo. Grandpa asked if I wanted some candy, and, of course, I said yes. Ice cream from a store was unheard of—no electricity and therefore no refrigerators, only ice boxes. One could make ice cream, of course, with salt and winter ice cut from local ponds and stored in sawdust in ice houses. But if you made ice cream in a hand-cranked freezer, you had to be there to eat it right away.

Grandma had packed a lunch of apples and sandwiches made with that home-baked bread. As we ate in the shade of a giant oak tree, allowing the horse to graze, we visited, talking to each other in Pennsylvania Dutch, the only language I knew then. Business done, Grandpa let me drive the buggy as we rode back toward Charm, on through Berlin, taking the back road which passed right behind my house, and on to Bunker Hill and Grandpa's farm. Grandpa told me stories about all the local people and about things my parents didn't think I'd understand. He talked about his parents, about his early married life, and about his struggles to become a respected citizen in the face of great odds between 1870 and 1900—man-to-man talk with a four-year-old boy! Usually, no one pays any attention to a four-year-old. Only covertly would one hear adult talk, and this was sometimes shocking!

During the time I lived with these wonderful people, I was treated to the thrill of watching my first baseball game. My Uncle John A.E. Miller, who worked for my father's company from time to time but spent most of his time running Aleck's farm, laid out a baseball diamond in the flat "bottom" to the east behind the large barn and straw shed. The men gathered round after Sunday dinner. My father played as an infielder, Uncle Rudy pitched, Uncle John was an outfielder, and Uncle Alvin played first base. Some of Aleck's farm hands and farmers nearby played other positions. The rousing game over, we came back to the house where they churned up a freezer full of ice cream with Uncle Alvin's ice cream freezer.

In those many contacts with all my relatives of the Alexander Miller extended family, I was privileged to develop a hearty sense of "family"—all this from the age of two, through high school, college, the university, my move west, and to this very day. Family became a very important influence in my life, leading to similar attitudes in my own burgeoning family which began to develop after my marriage in 1940.

Stephen Wengerd
December 16, 1853 - April 17, 1931

Elizabeth Shetler Wengerd
December 11, 1860 - January 12, 1948

Steffy and Betsey

Steffy and Betsey were married when they were young. He was born on a farm on the 16th of December, 1853, in the Sugar Creek Valley of Tuscarawas County. Betsey was born on a farm to the west in the Baltic area of Holmes County on the 11th of December, 1860.

When Steffy was a farmer working on his father's farm between 1870 and 1880 and then on his own farm until 1920, he learned timbering operations in his own woods, helped other farmers with technical know-how ranging from butchery to iron mongering, and bought and sold horses, cattle, and hogs. He had a sense of the dramatic in the way he wore his hat and his plain, somber Amish clothes. For many reasons, there never was much money available, so when he repainted his large barn, he painted only the side and end which could be seen from the country road that ran along the south side of the Home Place. So much for show!

Home Place

Steffy was a meticulous farmer, but more than that, he became an oil and grease salesman for John D. Rockefeller after the sale of the farm. Earlier, he had sold anvils and farm tools. Stories abound about his great strength as he carried 150 to 200 pound anvils from his spring wagon and set them up for blacksmiths and farmers.

My first memory of this couple was of our meeting at their farm near Doughty Valley, east of Beck's Mill in eastern Mechanic Township of Holmes County. They seemed formidable to me, though Steffy and Betsey were both only slightly over five and a half feet tall. She was a beautiful woman; he was a broad-shouldered man with black hair and a neatly trimmed black Amish beard. Betsey, blue-eyed, and Steffy, dark brown-eyed, were a handsome couple who were to suffer much tragedy and many moves during their years together. Their eldest son, Elmer, a highly religious young man, died of pneumonia at the age of 23. Two more sons, John and Levi, were born in the log cabin the family lived in before they built their new home in 1886. Everyone knew their beautiful new home as the Home Place. There my father, the baby of the family, was born in December of 1886. The boys left home—John to Pennsylvania, Levi to Wooster, and Allen to work for and live with Alexander Miller, my other grandfather who owned most of the acreage of Bunker Hill. As the boys left home, Stephen gave each one $1000. Unable to farm the Home Place by himself as he grew older, he sold the only real home he and Betsey ever owned. They lived with my parents for a year after 1920 and then moved to Pennsylvania to live in an extra house my Uncle John owned on a large farm on the east side of Negro Mountain. When he and Grandma moved to Pennsylvania, I was crushed, but I was to see them often as we took trips to visit them in 1922 and later. When Steffy died in 1931, Betsey continued to live with her son, John, until after his death in 1945. She then moved to my parents' home in Ohio where she stayed until she died on the 12th of January, 1948.

During the year Steffy and Betsey lived at my parents' home, he impressed me with his saving ways. I spent a lot of time with Grandfather Steffy Wengerd during that year. He had a great sense of humor, a sonorous voice and could be very impatient with my small boy peccadillos. During the summer of 1920, I would go with him into the fields and woods of our farm near Berlin to gather dock—he called it Indian tobacco—which he dried, cut up and mixed with his pipe tobacco as an extender. During the early winter when I was in first grade and walked through deep snow drifts to the school on the west side of the village, I'd come home complaining of snow getting up my pants legs and into my shoes. He insisted I should pull my socks up over the outside of my pant legs, which I thought looked terrible, so I didn't do it. Why I didn't have arctics, I don't know. Of course, I didn't really need to walk through those snow drifts overhanging the steep west bank of the road to school, but crashing down through such drifts and sliding out to the road was really fun. Those snow drifts drew a boy like flowers draw bees!

Steffy was a champion pig sticker. Once in the fall of 1920 during our butchering season, I saw him deftly kill a hog in our barn with one slice through the jugular vein. During that same butchering, I turned the crank of the sausage machine as he fed pig meat and fat into the top of the machine and expertly filled the hog casings with the raw sausage as it came out of the opening at the bottom of the machine. Steffy was widely sought on many farms to supervise hog butchering, and he helped devise smoke houses where ham and bacon were cured with oak smoke.

We made our first long trip in our Maxwell touring car when I was seven and my chubby little brother about four.

We started at 3 o'clock one summer morning. The speed limit was 35 mph, and there were no road signs, only colored stripes painted on telephone poles along what was later to become U.S. Route 40; red, white, and blue bands identified the Lincoln Highway, later U.S. 30. In the late afternoon, out of Uniontown, Pennsylvania, I saw the high rampart of Negro Mountain ahead, the only mountain I had ever seen. The visit lasted only a few days for business purposes—my father was the roofing wholesaler for Uncle John's roofing business which would eventually make him one of the first Amish millionaires. I have two memorable experiences of that trip: hearing Grandfather Steffy shout with delight as he swung on a swing tied high in an oak tree at the back of Uncle John's barn, and then watching him do forward flips from a standing position, landing on his feet every time. He was almost 70 years old then!

In 1925, Dad and Mom took us to Pennsylvania to visit our grandparents again and left us with them while Uncle John and Aunt Lydia went with our parents to Norfolk, Virginia—via Lancaster, Pennsylvania—visiting relatives and transacting business. During the ten days we were there, we were taken care of by Steffy and Betsey. We watched Steffy take care of his beehives in the front yard. One day he took his chew of tobacco out of his mouth, slapped it on a bee sting on his head and shouted bloody murder to Betsey as he ran into the house so she could help him pull the stinger out. We laughed, which made him very unhappy.

During those ten days we got very homesick. Grandma
was a terrible cook. Ants crawled around in the kitchen, breakfast was cold cereal, lunch was a coarse brown bread sandwich with a slab of fat smoked ham, and supper was some glop with no taste plus a cut of non-descript meat. Steffy spoiled Betsey rotten—cut up her meat for her, served

the supper—and, at times, did a better job of cooking. On rare occasions we had Cousin Menno and two Amish girl cousins to play with, but they were usually kept busy by a maid and an older cousin.

In 1929 or 1930, after several other trips to visit my grandparents, we drove in Dad's 1924 Hudson Brougham to Salisbury, Pennsylvania, for what turned out to be our last visit with those two interesting people, Steffy and Betsey. We took them for Betsey's first visit in 50 years to the farm of Sam Farmwald, the rich man who had married Betsey's sister, Amanda. They lived in the Lancaster area. We went to Gettysburg to visit the battlefield and the site of Lincoln's Gettysburg Address. Steffy was all excited but didn't stand the travel nearly as well as Betsey, who was tenacious and loved to go travelling. I was never to see Steffy again, and I won't forget the sad telephone call in April of 1931 when the Berlin line phone rang—we had a different phone for Millersburg—and Uncle John announced Steffy's death. Dad told his brother that he and our mother would come to Pennsylvania for the funeral. We boys were in school, so we were left at home in the care of Ida Lenhart, an older cousin who was my mother's maid. But Betsey lived on and on, an embittered but proud old lady.

When Steffy died, his will showed that he had a cow and some furniture, which went to Uncle John, and, wonder of wonders, $5000, a considerable sum then, which he had rat-holed so Betsey couldn't spend it. This he willed to their three sons under the proviso that they care for Betsey until she died—which she did in 1948 at the age of 88. Steffy was 78 when he died.

Betsey had been born into the Shetler family, well-to-do farmers near Baltic, Ohio. When she married Steffy, the Shetlers considered it a tragedy. She couldn't cook, hadn't been taught to keep a clean house, and spent

Steffy's money as fast as he could bring it home from his many money-earning activities. What happened to most of the money from the sale of the farm after he had given $1000 to each son, no one knew, even after Steffy died in 1931.

To Betsey's credit, years later, after World War II when she was living with my parents in Ohio, I realized she was the most intelligent woman—lazy though she was—that I would ever know. Steffy was of above average intelligence, didn't particularly like routine hard physical labor, and preferred to work at things that others didn't know how to do. As a salesman he had no peer. As I reached adulthood, I came to look upon these two grandparents in a different and more lenient light.

With their passing, all of their descendants realized that these two, Steffy and Betsey, were a most unlikely pair. Stories still abound among those who remember them. They talk about a rich young girl of 18 marrying a poor young man of 25 who was the son of a well-known but improvident Amish bishop who died far too early in life. Steffy and Betsey are buried in the lonely burial ground of the Conservative Mennonite Church near St. Paul, Pennsylvania.

Parents

Any person happily married is bound to have two sets of parents: his own and his wife's. It is reasonable to note that many people unhappily married or sequentially married to a number of spouses may not have had the privilege of having four parents. My mother was Elizabeth Miller, born on March 7, 1889. I vividly remember her since the age of one and a half. My father, Allen Stephen Wengerd, was born on December 14, 1886. I remember him from the time I was two years old. Obviously these memories hinge on specific incidents which I recalled for them both when I was old enough to realize that memories were important parts of one's personality.

My wife, Florence, had an equally amazing set of parents, but they were different in every way. Marie Porter was born on May 12, 1889, and I first met her in the fall of 1937, in Newton Centre, Massachusetts. Kirtley Fletcher Mather was born on February 13, 1888 and I knew him first at Harvard during the winter of 1936-37. In April of 1938, Mother Mather, as I later called this distinguished lady, decided that the graduate students in the Division of Geological Sciences should be invited to a party in honor of her oldest daughter, Florence, who was home from library science graduate studies at Columbia University. Her other daughters were also present—Judy was home from Bradford Junior College, and Jean, a pretty little sprite of eight and a half, attended Mason Grade School in Newton Centre. Dr. Mather dutifully invited the graduate students as directed, and some 20 students of his came to the party held in the basement rumpus room of their new home at 155 Homer Street. This house, a three story Georgian-style brick, was designed by Mother Mather, a Phi Beta Kappa graduate in mathematics and physics from Denison University. Dr. Mather, also a graduate of Denison, received his doctorate in geology from the University of Chicago, a pioneer university in that science in his day.

These details are recited as background in comparison to the milieu into which I moved as an occasional student at graduate school after graduating from the College of Wooster—I was the first member of my family to go to college.

My father went through the sixth grade—the highest grade level one could achieve at the one-room Oak Valley School where Harry Logsdon taught all six grades. Similarly, my mother went through the sixth grade at South Bunker Hill School, about a mile from her home on the Aleck Miller farm.

South Bunker Hill School

Even though my father went to school early and had, by law, to attend school until he was fourteen, he kept on going to school, and for the last two years, he helped Mr. Logsdon teach the lower grades! The earliest photograph we have of my father is a group photograph taken circa 1896 when he was ten years old. In that same photograph, Mr. Logsdon sits formally in the middle of the front row and the two older Wengerds, John and Levi, can also be seen. The eldest brother, Elmer, is not in the photograph, adhering to the strict Amish dictum against appearing in photographs.

My parents died in 1966—Father on April 20 and Mother on October 12. He was 79, and she was 77. Mother Mather died on September 17, 1971, and Kirtley died on May 7, 1978. Their lives spanned possibly the most fantastic changes in America, a time in which the horse and buggy moved aside for space travel and satellites in the heavens. The Mathers were dedicated Northern Baptists of Puritan-Pilgrim heritage; the Wengerds were life-long Mennonites, Germanic Swiss in origin.

From those two dissimilar backgrounds came two people—Florence and myself—who are both avid readers and perhaps slightly above average students with strong religious and family ties. My parents came from farm families, Florence's parents from small town to city families—where Mother Mather's father was a farmer and banker in Michigan, and Kirtley's father was a railway ticket agent in Chicago. In those times, between 1890 and 1980, in America, education was the key to progress, whether formal, obtained through reading and personal experiences, or both.

Me with my mother-in-law,
Marie Porter Mather
1969

Allen Stephen Wengerd
December 14, 1886 - April 20, 1966
Elizabeth Katherine Miller
March 7, 1889 - October 12, 1966

Al and Lizzie

When my parents were married in 1906, they moved to a farm near the Ashery north of Mount Hope to sharecrop for two years. A photograph of this handsome couple shows Lizzie to be a beautiful girl of 19, already a mother, standing beside my seated father dressed in coarse clothing and clod-hopper shoes, as though he had just come to the studio after plowing the back forty. My father hated farming; furthermore, he was on the "outs" with Aleck, his father-in-law. Aleck was already 54 years old, well-to-do, and eager to get out of the slate roofing business—a difficult and strenuous job for a person who was then considered to be an old man. He'd show the business to that young whipper-snapper son-in-law, Al. My father rose to the bait and bought the A. E. Miller Roofing Company in 1908.

World War I came along and the slate business boomed as tin and steel were in short supply. By 1918, my father and mother rebuilt the old oblong house they had bought in 1907, creating a handsome bungalow. Al and Lizzie had made plenty of money. She boarded from three to five roofers in our house between 1907 and 1930—cooking, doing laundry and cleaning house for those bachelors working for my father. They lived at our house so that they could work out of the roofing shop at home, first with horses and wagons and later with a Maxwell truck for greater mobility. My brother, Owen, went on the roof at age 10, as I was to do at age 11. Prior to World War I, the business was moved to a large new warehouse by the railroad in Millersburg, Ohio. Later, more trucks were bought and the business enlarged. All this time my father charged the A.S. Wengerd Slate Company for the cost of room and board for those roofers who lived at our house—a considerable amount of money then.

Dad at A.S. Wengerd Slate Co.

During the time between World War I—into which my father was drafted on November 12, 1918—and World War II, the business flourished, even during the years of the Depression. The company would take produce in trade for services and supplies when money was short. Father acquired three farms and residential lots in Florida, Ohio and Oklahoma. He made a long foray into south Texas in 1926 thinking he might move the business into the Rio Grande Valley. When he came back, he said simply, "Not enough people there. The ranches are too far apart."

Al was an innovator—a self-trained engineer who studied work flow and a salesman whom everyone but his competitors liked. He railed against Sears Roebuck for selling shoddy farm supplies, against the Farm Equity elevators for selling inferior feeds and fencing, and against the Farm Grange for not doing more to teach farmers how to farm more efficiently, even though the Amish used only horses to do farming. In fact, he was so vehement in these denunciations that his competitors hung the nickname "Sears Roebuck" on him in derision. All this before that hated Roosevelt "turned the country into a bunch of Socialists," as my father put it, and started the Soil Conservation Corps. He had to admit the SCC was one of two good things Roosevelt did—something the Grange should have been doing years before with the joint efforts of the farmers rather than the federal government, which "wasted money like it was going out of style."

All throughout the times beginning with the 1921 depression, through the better times of 1922 to 1928, and on through the Great Depression until 1939, my father experimented with a central kerosene stove and three new round brooders—round so chicks couldn't crowd into corners and suffocate. He designed two mass production chicken houses and set up an automatic switching system in his bedroom that would turn the lights on in the chicken houses at 4 o'clock in the morning. This woke up the chickens so they'd feed longer, lay more eggs, and allow my father to sleep until 5:30 AM. Selling chicken feeds bagged by the UBICo Milling Company in Cincinnati became popular as farmers realized that Al's 2,000 leghorn chickens were EACH laying up to 300 eggs per year. He began selling the chickens when they had laid for three years; the non-layers were culled twice per year. Culling was done at night. A chicken's laying bones under the tail had to be three or four fingers wide to be kept; two-finger-wide chickens were crated and sold for meat at a considerable profit to Mr. Levine from Canton.

The chickens put Mother in business. In an unbeatable arrangement, my father bought all the feed, supplied the electricity and paid the wages to workers (and you can guess who did most of the work—me!). But Mother got all the egg money free and clear to run the house. She was given no other money unless she asked for it—and she wasn't bashful about doing that.

My father's innovations are best typified by his view on electricity. When in 1920 other families went from kerosene lamps to gasoline lanterns or esoteric carbide systems (both dangerous), he set up a battery of 24 wet-cell batteries and a gasoline generator to make electricity using a central location in the washroom of our separate summer house. He electrified our house and put in modern plumbing while neighbors were still using kerosene lamps and backhouses! His henchman and general handy man, both in the 1917-1918 rebuilding of our home and the later electrification of the house, the barn and the chicken houses, was Uncle Rudy, my mother's youngest brother. When rural electrification came to Ohio, supplied by contract through Ohio Power Company, my father immediately changed over and sold his generator and batteries at a profit to a non-Amish family far out in the country beyond the electric lines.

Stories always seemed to pop up about my father's sagacity as a business man: how as a boy he once traded a gift knife through several exchanges up to a bicycle and, when he outgrew that, to a buggy and a horse and sleigh with sleigh bells and all! He sold his farms and turned his company over to my older and younger brothers in 1945, but he kept working for them until 1955. His pay was $150 per month, which was the cost of the company sold to my brothers. He had the reputation of being a rich man, and indeed, at one time, his estate was something over $150,000. Today, the inflated dollar equivalent would raise the price to well over a million dollars.

After retirement and when they could no longer travel easily, the window to Father and Mother's world was television. When Hudson cars got to be too low and cramped, he bought a large taller Packard. He said, "Anytime a car is so low it knocks my hat off, I'll get a bigger one!"—a car, that is. He always wore the finest suits, silk neckties,

Borcellini hats made in Italy, Dr. Adams shoes made in New England, starched white shirts—early with removable white celluloid collars and removable cuffs with sleeve holders (arm garters)—and the finest socks. All this until his weight topped 300 pounds as he stood five feet ten inches tall with shoulders seemingly a yard wide and a waist of 56 inches.

My parents at the A.E. Miller family reunion
1961

Neither he nor Mother were apt to spare much time to explain why they made us behave or do this or that. Dad made sure we learned by our mistakes with this blunt advice: "You know when the little bird shits in the nest, it has to sleep in it!" Orders were orders—do it, get going! But they were wise in telling and thanking us if we did well yet completely unafraid to let us know if we did something wrong. The last two great thrashings I got were, first, for hiding over at our neighbor's house as Mom called for me to come home; she had seen me dive under the porch! The second one was over my pesky little brother, who, against my advice, leaned far over the back of a porch swing, fell on his face and bit clear through his upper lip. I told Mother the truth and she still blamed me. "You ARE your brother's keeper," she said. As a little boy she never taught me how to tie my shoes; another boy in the first grade taught me. Challenged, she responded, "You're supposed to learn by watching!"

In June, after Dad's death in April, I went back to my college for the 30 year reunion of my class and stayed at our home. I heard Mom cough grievously in the night. I remembered that she had nearly died of edema in the lungs the week after Dad died, but Owen and a doctor saved her life. I also remembered that she had not cried at Dad's funeral. While I was back that June, I took her on a long day's motor trip to all her old haunts in Holmes, Wayne and Stark counties. As we visited I detected her deep sadness, yet knew she had a lot of fight left. She was almost exactly the age I am as I write this, and I asked her about the wild arguments she and Dad had so many times at the breakfast table where she always bested Dad with her logic. "How did you really get along with Dad?" I asked. With her very dark brown eyes afire, she said, "We got along fine. You know, Sherman, you can get used to anything but hanging!"

At certain risk of sounding maudlin, I recorded my thoughts when I again stayed alone in our home in Ohio for a night after Mom's funeral in the Berlin Mennonite Church. It was October 16. Here are those thoughts:

"As I walked around the house and saw all the things Mom loved and cared for, including her many plants and her carefully covered Singer sewing machine on which she had sewn up hundreds of sesquicentennial bonnets for the Berlin Sesquicentennial celebrated that summer. I also thought of the many large cauldrons of rivils she made for that same celebration. Rivils are a food her mother invented, and my mother, she was the expert! I looked up and felt the lamp shades she had made, the clean rugs, the clean mirrors, the

spotless kitchen, the many jars of canned goods she had put up in our cellar, all the dishes she loved, stacked in a glass-fronted dining room cabinet we called the 'schonck.' There even was the cup where she kept the little bit of money she required after Dad died. If ever my heart almost broke, it was this night."

Had I been inspired to call her on Tuesday after having telephoned her on Sunday evenings (as I did every Sunday the last three years they were alive), I could have assured her of our deep love for her and told her that Carol, my sister, and her husband, Dallas, had arrived safely to visit us in Albuquerque. It was not to be, for as we walked into our home after taking Carol and Dal out to dinner, our telephone was ringing. My brother, Owen, told us she had just died—would we come back right away? We rushed back to Ohio for the funeral. On the previous Sunday night, Owen was at Mom's house. Cryptically, she said, "Owen, I don't think I'll live to be 78." Owen joshed her, but she was lonely. That very evening she quit taking her blood pressure medicine, and by Tuesday she died of a massive stroke. As ministers say it, "The Lord took her quick, without pain." How do they know?

Dad always said that Mom was a proud woman. We knew her to be sensitive and sympathetic to all sufferers. She went to church "to help Jesus," as she put it, and she knew her position with God. Her nickname from little up, as the Pennsylvania Dutch put it, was always "Lizzie." She paid no attention to pettiness in others, nor was she ever jealous of anybody. She gossiped very little. Her sense of humor was fantastic, and she was an accomplished mimic with comical mannerisms that kept us laughing. She seldom showed her strong emotional make up. Her voice was bell-like, and she was one of those who filled a Mennonite church with heavenly voices as those Pennsylvania Dutch women made "a joyful noise unto the Lord."

In all those Sunday night calls to Mom and Dad, I was cheered by her voice and by Dad's enthusiasm. I literally thanked God for allowing them to live so long. But when Dad died, she was so lonely that she saw no good reason to stay on Earth. Dad had died within two days of their 60th wedding anniversary. Her single comment was "I just think it can't be that Dad is gone." The day she passed away she had cut the grass in the backyard and raked the leaves fallen from the great maples edging the front yard. Then she sat on the front porch and said as reverently as possible, but more than a little demandingly, "Lord, take me. I'm ready." How do I know that? She told that to my older brother, Owen, only minutes before he left to go to his house for supper the night she died. In June she had told me, "I'm ready to go. I'm not afraid to die!"

With parents like that, proud of the accomplishments of all their children and grandchildren; how could we possibly fail to succeed?

Kirtley and Marie

It is possible to look back on an association of 42 years with a man and gain some sense of his importance in one's life. Not just because he sired a girl who was to become my wife, but also because he married a person of uncommon splendor. Kirtley Fletcher Mather was born in Chicago, went to the University of Chicago for two years before his father, William Mather, decided his son should abandon that sinful place and go to a Christian college: Denison University in Ohio. It was there that he met Marie, an outstanding student of more directed energy than Kirtley's—at least she made the Phi Beta Kappa scholastic fraternity in her junior year in college, which he didn't; only later was he invited to join Phi Beta Kappa based on criteria other than academic achievements. They were married in 1912, and, after many moves, he finally earned his doctorate in geology at the University of Chicago in 1914.

My purpose here is to examine briefly the tremendous effects these two people had on my life. When I returned to Harvard in the fall of 1937, I was chastened by the fact that I had worked for six months in petroleum exploration as a jug hustler, a menial job on a seismographic crew in Oklahoma. Furthermore, it was obvious to me that I didn't know enough to become an exploration petroleum geologist! At Harvard, I changed advisors from Dr. Donald McLaughlin to Dr. Mather, because he had been a very successful petroleum geologist with world-wide experience. I took all the courses he taught, plus a number of special projects directed toward petroleum exploration, long before it struck me that he had beautiful daughters. I was too busy to think of girls amid 100 to 105 hours of courses, study, laboratory work, and independent study each week. Various jobs waiting tables and, later, laboratory teaching as an Austin Teaching Fellow for Dr. Mather left no time for extracurricular activity.

From 1937 to February 7, 1940 (when I passed the oral comprehensive doctoral examinations), I also worked on four special projects, reporting to him my progress each week. I noticed that he was extremely busy as a teacher, outside lecturer, a book reviewer for the Scientific Book Society of the AAAS, a fighter of stupid politicians and the press, and a man busy with innumerable other activities of which I had no knowledge. Each week I took exactly 15 minutes to report orally and ended each project with a comprehensive report, amply written and illustrated—all this while carrying the maximum course load an Austin teaching fellow was allowed to carry. Each Thanksgiving and at times during the Christmas and half-year break, Kirtley and Marie would invite to dinner graduate students who didn't go home. Toward the end of my next to last year, I noticed that the oldest daughter, having finished her studies at the School of Library Service at Columbia University, was living at home and working as a librarian in the Newton Public Library. Propinquity makes for marvelous eye-openers as one's formal academic work nears an end. The Saturday night after I had passed the comps, I asked Florence to marry me, and she accepted! The day after, I asked Kirtley, stating that I realized I came from a totally different background than Florence did. He said, "Florence's mother and I have noticed the great amount of attention you two are paying each other. Who cares about where you come from; I'm interested in where you're going. Of course you have our blessings. Now we'll have to see that you find a job, so I can drop some of that heavy life insurance I'm carrying."

That fazed me not one bit, and we both laughed.

Kirtley F. Mather (February 13, 1888 - May 7, 1978)
Marie Porter Mather (May 12, 1889 - September 17 1971)
On their wedding day, 1912

Only a few weeks later in the spring of 1940, the vice president of Shell Oil Company from Tulsa appeared in Cambridge on a recruiting drive with instructions from Roscoe E. Shutt, the Shell exploration manager, to look me up for an interview. I got the job! This led Kirtley to say, "I like that. You got your own job!"

Through the years, Kirtley became my best friend. Although we were diametrically opposed politically, we ranged through discussions of religion, politics, finances, investments, football, basketball, baseball, children—nothing was sacrosanct or out-of-bounds. About a year after Marie passed away, he moved into Pueblo Buenaventura, our large house in Albuquerque, and lived there until 1976 when I retired from teaching at the University of New Mexico.

From the first time I met Mrs. Mather, later known to all of us as Mother Mather, I was impressed with her good nature, love of humor, engaging conversation, and all-around femininity. She even enjoyed risque jokes and talking to me on the telephone. In fact, I thought so highly of the lady who became my mother-in-law that I often mentioned the probability that I married Florence because I admired her mother as much as I loved Florence.

Since the death of Kirtley and Marie, our children hark back to their many contacts with these unique grandparents, as they lived near us in Albuquerque. When our son, Tim, married in New York City, he asked his Grandpa Mather to be his best man. Reciprocally, when Kirtley remarried six years after Marie's death, his grandson, Tim, was Kirtley's best man. Of such are families bound together—a kind of immortality.

Dad as a Youth

He was born in December 1886, the youngest son of Stephen Moses and Elizabeth Shetler Wengerd, in a new house on a farm deep in the hill country of Holmes County. He had no memory of having nearly died when he fell into a very cold spring covered by a spring house on the Home Place. As a boy he was taught very early to do farm chores like chopping wood for the kitchen and living room stoves to milking cows, plowing fields, shocking wheat, pitching hay, and taking care of the horses with which my grandfather farmed the land.

He was named Allen Stephen. The first formal photograph we have of my father was taken in 1899 with an older brother named Levi in Millersburg. We have an earlier group photograph of the scholars at Oak Valley School, also known as the Mullet School. Although they never joined the Amish-Mennonite Church—called Old Order Amish today—as did the two older brothers, Elmer and John, their upbringing was as Amish boys with their clothes made by their mother, Betsey. Allen was strong, hard as nails and quick as a cat. His forte was his inventiveness, humor and storytelling ability. His memory was nothing short of phenomenal, but I am convinced that he was not above exaggerating a bit now and then in the anecdotes he told me when I was a boy.

His mother was a mediocre cook. The bread she baked was coarse and brown, and the meat was tough. The sugar they could afford was coarse and brown, but the butter was real, churned at home from cream gotten from cows Dad helped to care for. To his dying day, Dad would eat only white bread, white granulated sugar, real butter, and the finest of steaks and cheese, for as a boy he was always hungry. As the baby in a family of boys, his mother treated him like a girl. He rebelled and long before his father sold the farm, he took a job first as a farm hand and, at the age of 15, as a carpenter and roofer for Alexander Miller, who was to become his father-in-law.

The anecdotes he told me span the years of 1892 to 1903. As a little boy of six, he and his brother, Levi, were walking in the woods after felling trees with Steffy, who always carried a shotgun ready to shoot rabbits or pheasants for fresh meat. One day after several hours in the woods and a long walk home, Dad noticed that the 12 gauge shotgun was loaded and the hammer pulled back. For an hour or more, gun over his shoulder pointing backward, Steffy marched along through thick brush, over rocks, across fences with that shotgun cocked to shoot, the two boys directly behind their father. All the rest of my father's hunting life he insisted on teaching everyone, young and old alike, the danger of mishandling guns, based solely on that frightening experience.

When he was ten years old, Dad was at Beck's Mill, some miles west of the Home Place. He noticed some large trout in Sand Run, a clear water tributary of Doughty Creek which flows into the Killbuck River. He knocked on Mrs. Beck's door and asked for nails to make a fish spear out of a long branch and a cross piece. "Six nails, please. I'll pay you for them later," he said. The Becks ran a general store alongside the mill. She gave my father three nails saying, "That's all you need"—one nail fixed the cross piece, two nails through the cross piece to stab the fish. He marched to the deep stream, plunged the spear toward the fish from an angle, missed the fish, hit a rock, the cross piece turned, and he fell into Beck's Creek. Embarrassed by Mrs. Beck's laughter, he hied home to consider ways of catching those fish. He made a barbed fish hook, rigged a cork bobber, tied on a spent bullet for a weight, hooked it all up to a stout pole, went back to Beck's Creek, and caught the fish. When Mrs. Beck insisted that he give her some of the fish—after all it was Beck's Creek—he replied directly, "Nothing doing. You laughed when I fell in, and I fell in because you wouldn't give me enough nails to make a proper spear." My father became an expert fisherman with the finest equipment money could buy; it was a sport he enjoyed all of his life. He fished all over Ohio, the provinces of Canada, and at sea off the coast of Florida, those experiences being the basis of some of the most outrageous stories I ever heard any fisherman tell.

One day, felling trees in the dense hardwood forests on the farm, my grandfather chopped down a tree which fell on the head of Dad's brother, Levi, who was then in his early teens. Levi was knocked senseless, unconscious for a whole day. From that day to his death, Levi had mental problems, yet he was a highly successful paint salesman who raised a fine family. At times he was in mental sanitariums and at an advanced age died in the Massillon Mental Hospital.

At about age ten, my father built his own bicycle out of wood: solid wooden wheels, a wooden frame, a gunny sack on a piece of board for a seat, straight handle bars—a clever contraption. He had seen a bicycle in Millersburg, but there never was enough money to buy one. West of the Home Place, toward Oak Valley School, a long hill down the dirt road was the scene of Dad's first ride on his home-made bicycle. He had never ridden a bicycle, but he hopped on, gathering speed, shouting at the thrill of going faster than he had ever gone in a buggy. Part way down the hill his thoughts turned to "How am I going to stop?" He had forgotten to devise brakes, although he knew about brakes because he drove farm wagons with wooden brake shoes. Faster and faster he went, but at a bend in the hill road, he hit a fence, flew through the air in a somersault (called a *botzelbaum* in Pennsylvania Dutch), and landed in a hayfield on a haystack, the bicycle a wreck.

Undeterred, he was determined to get a real bicycle, and here the story becomes complicated. My father as a boy was known to be a shrewd trader. He told me he was given a pocket knife by his father. By various trades involving a pig, a broken-down spavined horse, and a spring wagon, he traded up to a second hand bicycle with real rubber tires and brakes. Proudly he rode that bicycle miles around, even into Millersburg, Charm, Berlin, Bunker Hill, Trail, Winesburg, Mount Hope, Saltillo and down to Beck's Mill. He was the only boy from an Amish family to own a bicycle! And he could ride it backwards as well!

When his father bought the farm and built the new house, they found themselves many miles from other Amish-Mennonite families in south-central Holmes County. Nonetheless, the Amish bishop, a stern forbidding man who, with an iron hand, ruled the Amish church to which Steffy and Betsey belonged, heard about Allen's bicycle. One day he appeared at the Home Place in his plain black buggy and told Stephen, "I want to see your boy, Allen, right now." Riding up on his bicycle came Allen, and the Bishop said, "Allen, a bicycle is a work of the Devil and not allowed for us plain people," whereupon he smashed the bicycle with rocks into a gnarled mass. Stunned, Dad and his father both reviled the Bishop—an unheard of action, as an Amish bishop was the moral law and had to be obeyed! For that one tragedy in the name of religion my father absolutely refused to join the Amish Church.

Another story came from a clandestine source. A

photograph of Levi and Allen holding a bunch of grapes and posing in a Millersburg photo studio suggests that it is true. On a Saturday sometime in 1899, perhaps near Christmas the year my father was 14 years old and Levi 16, these two tall sons of Steffy and Betsey went into a saloon to drink beer. Which one thought of it is immaterial. They had a formal portrait made showing the two, rather glazed of eye and serious. That photograph turned up after Levi and Allen had died as old men, so I never had an opportunity to ask my father about it. No doubt he would have come up with a long, rambling, humorous, shaggy dog story, perhaps unbelievable!

As boys, Steffy forced Levi and Allen to attend the Amish Church. Held at farms near Charm and Baltic, these church meetings were conducted on the threshing floors of bank barns. Benches were hauled from farm to farm every other Saturday for Sunday services preached in German by the Amish bishop. The benches had no backs and no cushions, and the preaching droned on and on for three hours. Reading the High German Bible, the Bishop required absolute silence as he explained the German text in stilted Pennsylvania Dutch. Everyone in such a congregation of Old Order Amish—at times numbering up to a hundred in size—was required to attend under threat of the *meidung*, a type of shunning which could lead to medieval-like ex-communication. Certain women were excused to spend the morning cooking dinner. After services, all gathered to eat a gigantic meal—that's the part my father liked best about Amish church services. The Amish never believed in having church houses, and the only music was hymns in German. All attended by horse and buggy with the more daring young unmarried Amish boys arriving in racy spring wagons.

The somber clothes of these people and the dreary church services were not to my father's liking. When he went to work for the liberal Mennonite, Alexander Miller, he joined the Mennonite Church. They were more modern people who could have telephones, cars, electricity, and educations beyond the eighth grade. Very soon after becoming a Mennonite, he began a personal campaign to influence the almost equally long-winded Mennonite preachers to reduce their sermons to less than an hour. Also Mennonite church was held in comfortable but plain church houses, where the benches had backs, though they were cushionless. Some of these churches had bells, and, until recently, the only music was singing.

My father showed his individuality early in life, and through a difficult childhood, he set the stage to become a self-actualizer who found ways to do things more efficiently. He became a talented and very strong side-hold wrestler. No one could throw him, which led to others trying all kinds of tricks to beat him at wrestling. As a practical joker, he had no peer, though he would be deeply irked if anyone pulled practical jokes on him. This was a game that trained him to become, at an early age, one of the leading businessman/contractors in Holmes and surrounding counties.

When I finally made it into the third grade, I came home one snowy day, complaining of the deep snow between our home and school, three-fourths of a mile away. My father launched into a story about how he walked three miles through deep snow drifts in south-central Holmes County to attend Oak Valley School. "In fact," he said, "the snow was so deep I had to walk on telephone wires to get through the drifts to school!" I was properly impressed, for I had never seen snow that deep. In 1930 when I learned to drive a car by myself, I drove the distance from where Dad was born to Oak Valley School, and it equalled one and a half miles. OK, he was right—to and from school was three miles, but I saw no tall telephone poles. Then I looked more closely for I knew my grandfather, then an Amishman, had no telephone. Strung along a low fence, not more than two feet off the ground, were two telephone wires! With this discovery I realized drifts could not have been more than two feet deep. I challenged Dad's story. His reply: "Well, I never told you how high the telephone wire was off the ground!" A lesson learned; he had not exaggerated—much!

Oak Valley School

National Air Races
1930

Travels with Father

Most of the trips I made while I was a youth were with both of my parents. Several later long trips with my father come back in memory mainly because he was a seasoned traveller who enjoyed his creature comforts and planned travel so as to stay utterly comfortable. All were by automobile, and it was during those trips that I really got to know this amazing man. Staying in the best hotels and motels, he always took along extra bright light bulbs—his comment to this being that "No hostelry has lights that are strong enough!" He travelled light, but often with his own pillow, and being as huge a man as he was, he insisted on a very soft bed. This was no doubt due to the rough corn husk mattresses of his childhood.

An early trip was in 1930 to the National Air Races in Chicago long before O'Hare Airport was built. The occasion also involved a national model airplane contest two days in duration during which I met the famous modeler and author, Joe Lucas, and other modelers of rubberband-powered model airplanes. Despite the Depression, my father took five days off from running his contracting business so he and I could spend two days attending the meet. We stayed at the Stevens Hotel—years later I was to give scientific papers at the same hotel during an annual national meeting of the American Association of Petroleum Geologists. We went to Curtiss-Reynolds Airfield early each day, watched the preparations, the flights, the crashes, and the speed with which modelers made repairs out of their portable work kits. It was a real learning experience for my own entry the following year into the International Wakefield Contests, representing Northeastern Ohio at the then Wright Field outside of Dayton. At the Chicago meet, I talked my father into paying for my ride in a giant Curtiss Condor, a four engine monster of a biplane giving rides to many of the modelers for $10.00 each. The air races by large planes were exciting, especially the fast pylon races with planes built by pilots who designed them and specially built them to compete in air races.

During all the years of the famous Cleveland Air Races in the late 20s and early 30s, we attended the event together, sometimes with Mother along. But the fascinating experiences at those races are another story because of the spectacular crashes witnessed and the famous pilots who participated.

During the Christmas vacation of the year 1932, my father had to make a business trip to St. Louis. Along was his good friend, Noble Uhl, who was to pick up a new car in the city known as the gateway to the Old West. Our trip was a fast one in my father's 1924 Hudson Brougham—I once drove the car at over 80 mph on a straightaway north of Wilmot, Ohio.

In 1955, with one of my father's ancient friends, the 85-year-old Noah Barrett, we drove to New York City. I was to lecture before the national AAPG on the newly discovered Paradox Basin oil fields in Utah. My father was then 69 years old and travelling with those two old friends who loved their bourbon was a trip of great humor. My father attended my lecture before several thousand geologists and averred that he enjoyed it: "If only I could have understood some of it." It was on that trip that he finally admitted that my becoming a geologist against his advice was a capital idea.

Far earlier, in both 1931 and 1932, my father took me to the Indianapolis Auto Races. We stayed in the same downtown hotel as Henry and Edsel Ford. We saw Eddie Arnold, the race driver who had won in 1930, go over the wall on the fourth turn of that infamous brick track in exactly the same spot both years. One year a Cummings Diesel race car, then carrying a mechanic with the driver, drove the whole race without stopping to refuel—a remarkable feat to advertise diesel engines for automobiles. As to Edsel Ford, who was staying on a floor above us at the hotel in Indianapolis, each evening whiskey bottles would rain down onto a lower roof after his loud raucous parties.

All in all, travels with Father were an important part of my education and the development of street-smarts. I have been able to pass along such knowledge to my own children—especially to my son, Tim, who went on several long trips by air with me to big cities. There are many special bonds that develop on such travels—a part of learning to cope with the dangers of an exciting world.

WE SELL
ALCOA
ALUMINUM
ROOFIN
Lowe Brothers
PAINTS
AMERICAN

Dad Wengerd

Foiling the Pickpocket

Few people have ever been as vehement about pickpockets as my father. He grew livid when he heard of anyone having his pockets picked. It was a badge of carelessness and dishonor to allow another human to mistreat anyone so. Besides, they worked in pairs, and Dad thought two-on-one was decidedly unfair. So, in his own way he declared war on those leeches on society. He was adamant that people around him hear his stories about foiling such dastardly thieves. It wasn't so much the money, he averred, as it was the loss of driver's license, credit cards, important scraps of papers, and the very idea of someone getting into someone else's pockets. So he fashioned a trap. He got an old wallet out of his pile of old wallets—he wore one out every year or so, carrying his wallet in his front left pants pocket. His wallet was always filled with more stuff than a woman's purse, plus as much as a thousand dollars in 20, 50 and 100 dollar bills. Into the old wallet he put cut strips of newspaper the size of dollar bills, placed a five dollar bill on each side of the inch-thick packet and carried it in his hip pocket so the bills showed.

Dad loved to go to horse races, boxing matches, baseball games, the Indy 500 Motor Race, any motor races within 200 miles of our home in Ohio, and into the teeming crowds of big cities. Street smart, lightning quick on his feet, and powerfully built with strong quick hands, he'd parade down the sidewalk waiting for a jostle. In Akron, Ohio, one evening after a wrestling match, he felt a tug at his right rear pocket, paid no attention to the accomplice who had bumped him, reached around faster than a snake's strike, and grabbed the arm that had his bogus wallet in its hand. In a wrestler's move, he snapped the arm, at the same time hitting the jostler with his ham-like left fist. He broke both the pickpocket's arm at the elbow and the jostler's jaw and left them writhing on the floor of the fight arena. Melting into the crowd—as well as a 300 pound man can who is somewhat less than six feet tall—Dad had foiled at least one duo of evil-doers.

As his sons grew up, he taught them how to hide money in shoes, around the upper parts of socks, in a hat, and in various parts of clothing. He always told them to carry the wallet in the front left pocket of a pair of pants with the left hand on the wallet, leaving the right hand free to swing a knock-out punch or grab a devious hand.

Years later, as people with me would have their pockets picked—with the attendant difficulties and unhandiness of cancelling credit cards and losing money and your driver's license—I would remember Dad's admonitions: Always walk in a devious line in a crowd; if anyone squirts shaving cream, chocolate sauce or ketchup on you making it look to be an accident, move away sharply and don't allow that person to help you clean off the stuff; in a crowd, talk to no one who asks directions and has someone nearby to jostle you; if you're being followed, duck into a doorway alcove ready to strike; if two people seem to be eyeing you and suddenly divide to surround you, stumble into one and cause a scene; and if you catch one, deal with him or her yourself, for the police will do nothing unless that person has your wallet in hand—a wallet in the hand of a thief is almost never retrieved once he disappears.

Any large city is a dangerous place at night, and this is particularly so in Mexico, Central America and South America. The most dangerous cities in daylight are Caracas, Rio de Janeiro, Colon-Cristobal, Panama City, Paris—with its gangs of teenage thieves—and New York City. The price of safety is alertness, and in many places it simply is not safe to walk. I was in a crowd of people in a subway one late night in New York after the Big Boat assemblage for the Statue of Liberty celebration. Jammed tightly, one of the riders said, "Any hand in my pocket had better be mine!" Everyone roared, for no pickpocket can ever be successful in a sardine can.

A particularly dangerous place is the new cathedral in Mexico City. A visit to that place is a must for tourists—and a real haven for well-dressed pickpockets. On one visit, I saw two of them scanning the crowd and watching me. I circled around behind them, and as they turned and saw me coming straight at them, my right hand menacingly bulging in my right coat pocket, they fled. Pickpockets are the world's greatest cowards.

My wife and I have walked in many dangerous places but walk as if not associated with each other. I walk behind; if her purse is not over her shoulder and under her coat, but is carried in her hand or over her arm, she walks inside away from the curb as I follow closely behind. In South and Central American cities, the incidence of purse snatching by passing bicyclists on motor scooters and motorcycles is high. It is standard practice for street-wise travelers to wear no jewelry in dangerous areas—that means almost any foreign city and many parts of larger American cities, especially in the evening and at night. Few of us are as large, impressively strong- looking or as alert as my father was, but I have learned his lessons well: Be watchful and let no one distract you!

Dad (second from the right) with
business associates

Dad's Friends

My father's storytelling led to his having many friends who enjoyed his company. As is so often the case, early in his life his friends were his cousins, his brothers-in-law, and the men who worked for him. Even earlier, his friends were men who worked with him when he was a roofer for his father-in-law, Alexander E. Miller. As Dad prospered, his friendships shifted to local doctors, lawyers, a funeral director, salesmen, and several people who held important positions in the Republic Steel Company of Canton. These became his life-long friends and, along with his voracious appetite for reading, inspired him to become a self-educated man and a self-trained engineer.

One salesman, Irwin Friend, was a friend and participant in many escapades with Dad. Friend was a member of the prominent Kentucky family that took in an orphan named "Happy" Chandler, later the governor of Kentucky and the United States Baseball Commissioner. Friend had a death-like pasty chalk complexion which led to an astounding practical joke Dad and Friend played on a mutual buddy, Ott Elliott, a funeral director. They fixed Friend up and laid him out like a corpse. To the funeral director's chagrin, the corpse sat up and cursed a blue streak. Paul Franks and Fred Hoffman were also friends, and Dad admired them because they kept the company truck and Dad's cars in top mechanical shape. Paul was manager of the American Legion baseball team on which I played from 1930 to 1931. Noble Uhl was a longtime friend who was always in debt but great at telling stories. From 1910 on he helped Dad become a polished gentleman in speech, dress and attitude. They were friends until Dad died in 1966. Noble lived into his 90s, dying in 1989.

In Dad's last 20 years of life, Noah Garrett, an old man in his 80s, became a good friend and travelling companion to Dad. Their chief activity was attending horse races and surrey races where they tried to outdo each other in winning bets. Both told great lies about their betting prowesses. Coach L.C. Boles and Stanley Welty of the College of Wooster were also Dad's friends. Boles was the football coach and Stan was one of Wooster's best-known football players; he was later a salesman who called on Dad to sell him fencing and other equipment. Hank Critchfield, a lawyer in Wooster and a former professional football player for the Canton Bulldogs, was a valued friend. Dan Swickard, a salesman for UBICo Milling Company in Cincinnati, called on Dad's company to sell bagged feeds to farmers.

Louis Bromfield, owner of Malabar Farm located south of Mansfield, Ohio, and a nationally-known author, was a regular client of Dad's and became a good friend. Paul Hummel, the butcher's son and, later, himself the owner of a meat store, was a friend much younger than Dad. Later, Paul became a highly successful insurance agent and, finally, a Mennonite minister. Two of Dad's very best friends were John Lennon and Ward Lanning. The stories of Dad's adventures with Lennon, a life insurance salesman with Mutual of New York, are legendary and are still told years after both of them died. The three once bought a new Ford, drove it to Alaska on the Alcan Highway, sold it at an exorbitant price, and flew home first class—the first and only time Dad ever flew in an airplane.

As many friends as he had, Dad once said, "You can have too many friends. They can get you into trouble. They cost you money, and you can't depend on them if you go broke." There were many reasons for that surprising statement from a man who truly enjoyed his friends. One of Dad's very earliest friends during the time he owned a two-cylinder Maxwell (circa 1910) was a man named Augsberger. In the early days—in fact from the age of 13 on—Dad loved beer. One wintery night he and Augsberger were drinking beer in Fredericksburg, and Augsberger backed up onto a stump forcing both wheels and rims off the ground. Racing the engine did no good, so they had to get an Amishman with a team to pull the car off the stump. That story haunted Dad a long time, so when my older brother, Owen, wanted to taunt Dad, that story did it.

I have made many tapes of people who knew my father well, and I realize that he was a complex man. His story, his life and his philosophy merit a book, but before that, this separate essay is an attempt to fathom how a man could come so far. Ohio was a frontier between 1886 and 1966. Dad was born into an Old Order Amish-Mennonite family that lived on a farm far removed from villages, towns, and cities, and he had only a sixth grade education from a one-room country school. He had few educational opportunities, yet he was widely known and well-liked throughout Ohio.

Dad with a
business associate

A Mouse in the Soup

I wasn't there when it happened but I heard about it from several sources. One of my father's almost impossible practical jokes is still told by older men when I ask them to tell me their memories of Al. On Main Street in Millersburg, on the south side of Jackson Street, the Thomas Restaurant was THE hangout for the noon lunch crowd between World Wars I and II in that town of 2,500 souls. Old John Thomas and his wife, Nettie, did the cooking, waited on the counter and cleared up after breakfast, lunch and early dinner. Old John walked with a limp from a wound suffered in the Great War in France. He and Nettie had two children, one of whom was a sailor in World War II in the Pacific; he was always afraid to die and finally curled up and shot himself as an "old," young man in the County Home.

The long counter, running north to south in the narrow room, was flanked on the east by some small tables at which two could sit if the counter was full—which it was every lunch time because Old John and Nettie made the best soup and steak and gravy sandwiches in the whole county, maybe even the whole state of Ohio. If the counter was not full and you tried to sit at a table, John would curse and tell you to get your ass over to the counter: "Be darned if I'll walk all around the south end of that long counter to serve you!" Nettie agreed, and you moved to the counter. If it was full, they'd have someone at the counter hand your meal across to your table—no wasted motion!

Old John was the ultimate curmudgeon. He was voluble in his cursing yet comical. He could tell dirty stories and sing the "Mademoiselle from Armentiéres" with all its verses, for he had been there. His raunchy voice would rise as he sang, "I'd stand her up against the wall, and push it to her balls and all." Women didn't much like to eat there, except for the lady who ran the local hotel—a refined woman who knew some raunchy songs herself and enjoyed eating with the business people and workmen who ate there every day.

Old John Thomas was the soup maker—he wouldn't let Nettie touch it. No one was allowed to see what he put in it. The kettle was apparently never cleaned out, or if it was, it didn't look like it. Into that soup went leftover gravy, fat trimmed off of steak, steak left on diners' plates and chipped into chunks, leftover vegetables, dollops of dough which cooked into small dumplings, black pepper, a touch of salt, and God only knows what else. He made the soup each morning before breakfast so it could simmer until the noon crowd arrived. Everyone had a large bowl of John's famous soup. As a small boy, I would go to the restaurant with my father when I accompanied him to the warehouse for the day—that way, I wouldn't be underfoot at home. I loved that soup. I wondered why my mother couldn't make such heavenly soup!

Old John was easy to josh; he would rise to the bait and give back better than he got. More than once pandemonium would break out as some customer would belittle the soup or bitch about the size of the plate of steak and gravy with mashed potatoes—even if the plate was overflowing! John's temper was a thing to behold—I learned some of my choicest cuss words from Old John. Everyone badgered him to tell them what he put in the soup to make it taste so good. The answer always was "None of your darn business. Eat, pay up and get out of here!" He'd shout to a particularly offensive questioner, "And shut the darn door when you leave."

One day Dad found a dead mouse in the soup. He always ate early, so the soup cauldron was full. Dad held up the mouse shaking the soup off of it. People quit eating their soup, and Old John blew his top as he carried the whole kettle of soup out to the alley for the stray cats and dogs to lap up. The whole restaurant was in an uproar: A MOUSE IN THE SOUP!—unheard of! Old John was fit to be tied, but he had lambasted Dad unfairly once too often.

That morning the warehouse cat had brought a dead mouse into the office at Dad's place of business at the foot of Mad Anthony Street by the railroad depot. Dad put the mouse in his coat pocket—as a prominent business man at much the same level as the local banker, he always wore a suit coat and tie to work when he ordered roofing crews to work each morning.

Going to the restaurant for lunch as usual, he ordered a bowl of soup and slipped the mouse into the soup. He "discovered" it to his faked consternation and held it up for all to see. That's when all hell broke loose. Raving and ranting, Old John paced back and forth, cursing, muttering, crying, and cursing again. "How could a mouse get into the darn soup?" After a while things settled down and my father, the good Mennonite that he was, quit laughing. Old John smelled a rat (not to pun about such a serious matter). "You put the darn mouse in the soup, didn't you, you bastard. I'll shoot you if I can find my gun!" Dad was afraid of nobody, and he knew Old John said some things he didn't mean when he was angry—for they had known each other since about 1910.

That same lunch hour Dad apologized so all could hear and laid a ten dollar bill on the counter—soup was 10 cents for a big bowl in those days. Old John was partly mollified, but forever after the incident when Dad came in for lunch, Old John would sidle up to Dad in his grumpy fashion and whisper, "You son-of-a-bitch had better not have a mouse in your pocket!"

With fellow geologists
overlooking Goosenecks of the San Juan River
Summer 1951

Vindiction and the Human Spirit

It is difficult to philosophize about vindiction in the light of Biblical New Testament admonitions about its use in human affairs. To consider hell which doesn't really exist, or purgatory which is equally as nebulous, as a fitting place to see wrong doers get their comeuppance is alien to the human spirit that endures human-created misfortune. In the clear lessons from the study of evolution, one finds abundant evidence that animals suffer torture and death for behavior inimical to the good of other individuals or groups of individuals. One need only study starlings, crows and penguins to see vindictive nature righting a wrong perpetuated by a wayward individual. For what reason, then, should any human suffer from heinous and not so heinous behavior inflicted by others?

To lean toward the Old Testament dictum of an eye-for-an-eye and a tooth-for-a-tooth creates a good feeling toward ancient Hebrews who knew full well the concepts of cause and effect: Do wrong and you get punished! The teachings of Christ may be in error on just that one statement: "Turn the other cheek." It is heresy, indeed, to say that if corrections aren't made forcefully in this life, then evil will eventually rule this world. Benjamin Franklin once said he fully intended to talk to God when he went to heaven, but wasn't too sure that he would find somebody called Jesus, Son of God. He wasn't banking on encountering Jesus to answer his questions as to why certain things were allowed to happen on Earth. It is possible that I have misread old Ben's approach, but I do know that he was the ultimate diplomat in applying vindiction unnoticed by the wrong-doer. Therein may be the solution to the strong admonition to "Turn the other cheek." Don't let the miscreant know he's going to catch hell on Earth; just turn the other cheek and get him when he isn't looking.

You may wonder how it is possible that a boy raised in a religious Mennonite Swiss home in Ohio could come to such a conclusion. My mother believed firmly that evil-doers would go to hell and doers of good would go to heaven. My father, the strong Mennonite that he was, felt that anyone who willfully wronged someone else—particularly himself—should pay for it now, immediately, not later, and certainly not for some ethereal deity who didn't seem to care to take it into His own hands. My father saw too many bad people become successful without stricken consciences to hinder them in continuing their misdeeds. One simple example will suffice—quite a gentle one. My father's company was expert in installing complex hay-track equipment in barns for farmers across a wide section of the state of Ohio. This equipment, with slings, harpoons and a complex system of manila ropes, picks hay off a giant load on a wagon drawn onto the floor of a bank barn by tractor or a team of horses. The hay is lifted to the peak of the barn and, by manipulation of the ropes, is carried on a track over the barn's haymow and released by a line-activated trigger. Then the harpoon or sling travels along the track back over the hay wagon for another load and so on. It was the best equipment money could buy; it never—or almost never—broke down, and if it ever did, he repaired it at no cost, even if the farmer was at fault for mishandling it.

A neighbor of ours bought such a hay-track system from Sears, Roebuck at a low price and had a local yokel install it—a man whom my father had fired for being incompetent. I shall never forget one evening during haying season. This neighbor appeared at our door and asked my father's trained experts to come out right away to fix trackage he had refused to buy from my father. His hay was dry but the rainy season was to be here shortly. It is critical to the proper drying of hay that wet hay not be put in a barn because of the risk of spontaneous combustion which causes barn fires. In a most diplomatic way, my father told him, "We are far too busy. Sorry. Why don't you get the man who put the cheap Sears, Roebuck track into the barn in the first place?" This was a mild form of vindiction.

The scene turns to smart-aleck farmers who had my father's competitors roof their barns with cheap roofing, incorrectly lapped and nailed by local men not expert in the installation of corrugated roofing. When their roofs leaked resulting in the possible spoilage of stored hay and grain, these farmers knew of our company's expertise in roof installation and requested that my father send crews to fix their roofs. To my knowledge, such roofs weren't fixable, and it was simpler to tear the roof off and put on good steel—by experts. My father would never just repair a cheap roof. Such vindiction never brought discredit to my father. Everyone knew that if lightning struck a barn and it burned, my father would be there to sell the distraught farmer a new guaranteed roof put on at no cost for labor. His profit on the roofing itself was negligible. I know, because I was a roofer and knew the cost of labor and of the roofing.

It was by such examples that my father taught hosts of farmers not to cross him by buying cheap equipment. In the world of business, vindiction, overt or covert, is a tool with tremendous power for rightness and efficiency. My father's world was filled with dealings with shrewd and clever scoundrels, experts in short cuts and double dealing. In few cases was he ever bested, as it became well known that he would give a farmer the shirt off his back if tragedy hit; but if they crossed him, he would somehow get their shirts without their knowing it. To my knowledge, he never hurt an honest, forthright man—a credit few humans can claim.

The stories he told me may have had an exaggeration or two here and there—for he was a superb storyteller—but they revealed exemplary ways through which a great human spirit could use vindiction to make the world a better and more logical place.

Dad at his desk
which is now in Albuquerque

The Bell

A large black iron farm bell hangs in a special bell arch at our haciendita in Albuquerque. It's not really an arch but a wall-like projection made of slump block with a massive wooden lintel. Two thick spruce uprights with metal gimbal sockets hold the bell so it can be rung with a pull rope. This bell was cast of Appalachian Mountain iron early in the 19th century by the Eagle Bell Company of Lebanon, Pennsylvania. One of many bells cast in the long history of that company, this melodious bell has a dimension of over 18 inches, is about one inch thick, and stands some 16 inches tall. The top is flat with a diameter of eight inches. Anchored on the top is the bell spindle in sort of a flat upside down "U" shape. At the ends are the metal bearings that fit into the gimbal sockets sunk into the spruce uprights.

The bell has been very special to the Wengerd family for possibly five generations since the time of Christian Wenger, who came to America as a boy in 1801. Typical of Pennsylvania Germanic Swiss farmers who always had bells, this bell has hung in at least five bell towers on farms owned by the Wengers and Wengerds, including ones in Brothers Valley of Somerset and Cambria counties in Pennsylvania and at Christian's and Moses' homes on farms in the Sugar Creek Valley of Tuscarawas County. It also hung in central Holmes County at Stephen's home where my father was born, and in eastern Holmes County where I was born. And now it resides in Bernalillo County, New Mexico, my present home since 1947.

To a non-Swiss family, the bell may simply be a

decoration, but to the Amish-Mennonites, bells were a necessity. They were used to call men to meals from the fields and barns, to peal in times of disasters and accidents when help was needed and to ring on Sundays as a "Joyful noise unto the Lord."

There are many stories about the Wengerd bell; I was sure, however, that my father, the storyteller, embellished the truth somewhat in telling them. He told me he thought the bell was cast early in the 1840s and bought new or used by Christian. The bell was probably hauled to Ohio by wagon after 1850 when Moses accompanied Christian's family on their move to Ohio. Moses and his father travelled along the old Indian trails and wagon roads between Somerset County and the Sugar Creek Valley. I suspect that these stories have been handed down from generation to generation as we do not have specific dates for such events. In those pioneer times, education in Ohio's hinterlands was sketchy at best. No letters—if any were written at all—survive; in fact it's doubtful any mail service existed between the Amish communities of Pennsylvania and Ohio. But one thing is certain, for I have the large German bible to prove it: my ancestors were literate. They could read and write German, speak English and were avid students of the Bible printed in Germanic script. In that Bible—which will eventually rest in the library of the Somerset County Historical Society—is inscribed the lineage of my family, from Christian's time to my father's generation, written in German. However, no written mention of the Wengerd bell is known.

From its location at the birthplace of my grandfather, Stephen, son of Moses, in Sugar Creek Valley, since before the Civil War, the trail of this bell is well-known. I first saw the bell in 1918 hanging at Stephen's farm. It tolled the end of World War I and could be heard all over the wooded hills of south-central Holmes County.

Every bell has its own timbre; the ringing is characteristic and identifiable not only by the sound and the rapidity of tolling but by the direction of sound travel. If tolled slowly, it was a call to men and women working in the fields; if tolled moderately fast (as though the toller was already late getting his buggy ready to go to church), it was Sunday morning; if tolled rapidly, the signal was a call for help NOW!. Such a disaster occurred at the Home Place in 1888. My father was a year and a half old. South of the house was a spring house with a deep and very cold spring of clear water. Betsey, Stephen's wife and my father's mother, was busy in the kitchen and missed Allen. Calling all over, she finally thought to look in the spring house and found my father's body submerged in the very cold water. She alerted Stephen to the situation, and he tolled the bell before running to the spring house, as my grandmother pulled my father out of the spring. They kept tolling the bell as they quite accidentally laid the boy's head down-slope adjacent to the spring. Neighbors rode up on horses. No one knew what to do. Suddenly they heard a loud bellow from the apparently drowned boy as he sat up and cried. They all believed it to be a miracle. The cold water had run out of his lungs. No one knew how long he had been in the spring, but he lived.

That story was told to me by my grandmother, Betsey, and my grandfather, Steffy, when they lived with us in Berlin during 1920-21. Needless to say, my father lived a long active life, a strong man of fantastic vitality. He died near the age of 80.

When the Home Place was sold, my father, the youngest of the four boys, took the 80 pound iron bell and mounted it on the south gable of the roof on the summer house adjacent to our home near Berlin. In 1950, long before the death of my parents, my brother boxed up the Wengerd Bell and sent it by rail freight to me in New Mexico. I built a true bell arch and hung the bell at our hacienda, Pueblo Buenaventura, in Albuquerque. We rang it to call our children home for meals. The bell tone was recognized all over Sunset Mesa near the university.

We sold Pueblo Buenaventura because the large, rambling, adobe-style 13 room house was too large after our four children grew up. We moved to a small house which we built adjacent to Pueblo Buenventura, and I built the special square "arch" to hang that historic bell. Today my grandchildren love to ring the bell, a legacy of a pioneer Pennsylvania Dutch Mennonite family whose ancestors began the migration to America between 1727 and 1801.

Pueblo Buenaventura

Crack of the Bat

Baseball is a game of superb skill involving a certain amount of danger. Some say it's more fun to play than to watch. People who've never played the game consider it slow. Many recognize in its apparent slowness the guile of a game filled with cat and mouse tricks, feints and all kinds of efforts to mislead the opposition, both on offense and defense. Watch a third base coach giving signals; observe a batter, with averted eyes, trying to see where the catcher is placing his glove a split second before the pitch; see the second baseman moving to his left as the shortstop tries to move behind a runner off second to take a throw from the catcher; and notice the runner on second relaying the catcher's signals to the batter—the oldest trick of all. The hidden ball at first base when the pitcher smacks his glove as though the first baseman has returned the ball; the balk by a pitcher to catch a runner off stride; the shortstop or second baseman leaping high into the air to catch an uncatchable line drive, making the runner think he can always catch those, so the runner hesitates; a fielder faking a throw to one base and letting it fly behind a runner for a play; the fake bunt; the hit-and-run; the run-and- hit—you name it, a good ball player will try anything to confuse the opposition. That makes the game a supreme game of wits.

It's also a game of rare humor. Why do you suppose such men as Casey Stengel, Yogi Berra, Joe Garagiola, Gaylord Perry, and many others became famous for their apparently inane statements about the game? Casey once said, "Good pitching will always beat good hitting and vice versa." Such pronouncements have led to the formation of "Stengleze," a language all its own. No wonder! Willie Keeler, an old-time player inducted into the Hall of Fame, observed, "Hit 'em where they ain't." Back between 1926 and 1932, we would go to Cleveland to watch the Indians play the New York Yankees, not just because Babe Ruth, with his high jinks on the field, and Lou Gehrig were playing, but because two of baseball's greatest clowns, Al Schacht and Nick Altrock, were coaches at the corners. Those men were former expert baseball players whose humorous antics were better than any by comedians at a burlesque show. Lou Gehrig, whose real name was Ludwig Heinrich Gehrig, once said, "When two men ride a horse, one must ride in front of the other"—a momentous statement by the Iron-man of baseball. He was as fine a man as the Babe was a devil-may-care wastrel, yet they were friends.

The two saddest moments I can remember in all of baseball lore occurred when Lou Gehrig on July 4th, 1939, and Babe Ruth on April 27, 1947, both ill, were accorded the accolades and plaudits of great crowds of admirers at special honoring ceremonies in Yankee Stadium.

The humorous observations go on and on—it's the sports writers and television announcers who keep the stories going. The baseballer's credo, believed by many of the old-timers like Rogers Hornsby, was "Kick their butts, drink their beer and steal their wives." Don Drysdale, one of the great pitchers of all-time, thought "Life is full of loopholes." He was often blamed for throwing "spitters" (or spit balls), and he once demonstrated all the techniques involved. Baseball players are the greatest superstitionists and the penultimate jokesters; they'll sit in a dugout, down two to three in the 11th inning, with their caps on backwards or inside out attempting to change the team's luck.

A player named Woford once said, "Ninety percent of

the game of baseball is half mental." Compare that to Yogi Berra's "Fifty percent of the game of baseball is ninety percent mental"—such switch jokes must be copies turned around. Yogi's include: "When it goes up you sell. When it goes down you don't buy"; "Avoid intersections and buy condominiums"; "You can see a lot by looking"; and speaking of Fenway Park's "Green Monster" which limits homeruns to left field, he commented that "It sure gets late awful early out there." And about a restaurant he didn't like he said, "Nobody goes there anymore, because it's too crowded." Joe Garagiola has often mentioned Yogi's real name on TV as Lawrence Peter Berra. Both of those exuberant Italians typify the humor in baseball. Satchel Paige's sage observation was to "Throw strikes. The plate don't move."

The game of baseball was my favorite sport in college, but our coach thought I was a flash in the pan, so I didn't get to play as much as I would have liked. I played on American Legion teams and in high school after we gave up softball as a sissy game. By then Berlin High School could afford to buy three bats and four balls and to find another high school not far away to play against. We had no uniforms until I played on the Berlin town team whenever I was home between 1930 and 1937. Even though I was light weight, short, not especially fast, had small hands, but was not quite as bowlegged as Honus Wagner, I still played a creditable center field and pitched a few games in high school, college and on the town team.

The high point of my non-illustrious career in baseball was the day I hit three home runs against Kenyon College as a junior member of the College of Wooster team. Kenyon's pitcher had been drinking beer. Of course, Bob Fuhrman, a senior, hit four home runs in the same game, so, as usual, I was second best. The greatest thrillers occurred in the spring of 1935 in the ninth inning of our college home game against hated rival Oberlin College. With two outs, and Vern Dodez, our catcher, on second base, Coach Art Murray put me in to bat, and I hit the ball deep into left center field as Vern raced home. We won not only the game but our college league championship too. To my surprise there are people still alive who remember that hit! In the late summer of 1935—just back home from summer field school with the Oberlin College field camp in Highland County, Virginia, and newly recovered from a session of severe bronchial pneumonia—I joined our Berlin town team to play the Shreve town team in Lakeland, Ohio. Playing left field, I came up to bat three times against Shreve's imported pitcher, Denny Galehouse of Doylestown, a towering left-hander who had played for years in the major leagues. I struck out twice and flied out to left field. Guess who was watching in the stands: my college coach, Art Murray! By spring of my senior year in college, geology was my consuming interest. The coach left me off the travelling team, and I threw in the towel as a baseball player. My father was disgusted, as his ambition for me was to play in the major leagues for the Cleveland Indians.

So ended my inauspicious seven-year career as an amateur baseball player. But today the game is still my favorite because it is tricky and filled with humorous gambits.

Baseball in Cleveland

The Aviators

It should come as no surprise whatsoever that I believe airplane throttle jockeys to be very special people. Humans, since the first uprightness of their australopithecine ancestors, have navigated the surface of the earth in essentially two dimensions of space. Since the days of the mythological Icarus, and Lillianthal in particular, a new breed of human—or an old breed with a new ability—was taken to orientation and movement in three dimensions. It is only within the last century that great numbers have learned to fly in contraptions as variegated as the human condition itself. It is not unusual that I should wonder at this ability since I learned to pilot airplanes within 28 years of the first flight by the Wright Brothers in 1906. More people can fly airplanes today than all the people who had ever flown airplanes.

My very first memory in life was of an airplane—a biplane, probably a Curtis JN4 "Jennie"—flying out from behind a cherry tree on the edge of our cabbage patch. I tried to grab it in my hand, and when I couldn't, I ran to my mother and complained, "*Verwas kannich es net grighe*?" It was early one midsummer morning in the year 1916, and I was wearing a dress. I was only one and a half years old, just some six months after my left leg was broken when my mother had fallen down the cellar steps landing on top of me—an event I don't remember. I could walk, speak Pennsylvania German, but my eyes could not focus adequately, so I thought, "Why can't I grab that thing" in the sky? It flew from southeast to northwest right over our town of Berlin and at least a fourth of a mile away from where I was standing. Such are first memories in guiding the intense interests that have never left me.

In the very early days of the 1920s, there were five pilots in Holmes County. The first two were Truman Bucklew of Killbuck who flew a Standard JN1 biplane and Howard Shoup of Winesburg who flew a JN4 Curtiss Jenny. The other three were Bob Marshall, Mutt Lichty and Dutch Geib of Millersburg. Lichty was injured in a spectacular crash on railroad tracks west of Millersburg. His JN4 went into an uncontrollable spin. He died of head injuries sometime after the crash. Below is a photo of the wreckage which was almost unidentifiable as an airplane. Lindberg made his epochal flight across the Atlantic to Paris in 1927. In the fall of the same year, Dad and Mom took me to Akron where an air show was held in connection with the dedication of the hangar in which the dirigible *Akron* was housed. Rides were being given in a Ford Trimotor for $5.00. I badgered Dad into letting me take a ride—my first. He relented, saying, "At least it has three motors." Much later I was to learn that it was the Trimotor owned and piloted by a man named Johnson, the nervy pilot who had beefed up the plane to do stunts such as rolls, loops, wing-overs, and Immelmanns—unheard of for a multi-engine airplane. In 1930 I actually met Wilber Stout, designer of the Ford Trimotor, at the Chicago Air Races!

Earlier than 1927—possibly as early as 1921—my family took my two-year-old little brother and me to an air show featuring a squadron of shiny new JN4 Jennies flown by Army Signal Corps pilots. Located near the town of Stow, northeast of Akron, the grass field was surrounded by thousands of people who parked their touring cars to watch these daredevil World War I Army pilots take off and land in "V" formations. One pilot got caught by a gust on takeoff and scraped the lower right wing on the grass to the oohs and aahs of the enthralled crowd, but he recovered without an accident. Seeing those airplanes fly and hearing their watercooled OX5 engines roar was one of the greatest thrills of my life. I vowed to Dad and Mom, "Someday I'll learn to fly airplanes." They pooh-poohed that idea with the comment that "Airplanes aren't here to stay. They're too dangerous."

Thirteen years after seeing those Army pilots, I started taking flying lessons and learned the pilots' curse: "Flying consists of hours and hours of boredom, punctuated by moments of sheer terror." In the period between 1934 and 1974, I had nine forced landings; fortunately, I did all without harming the airplane or myself. Air crewmen during World War II spoke of "chewing washers" when in dangerous situations—an apt expression.

Between that first flight in 1927 and my first flying lesson in 1934, I took many three dollar rides in barnstorming airplanes, flying out of hay fields, grass strips and hilly fields in central Ohio, to the vast consternation of my parents. With great reluctance Dad signed my application for a student pilot's license in August of 1934, and the dream of flying began to be realized. My instructor, Roy Poorman, held flight license #4773 and mechanics A & E license #4221, both very early licenses. I soloed after only five hours and 39 minutes of dual at $6.00 an hour. On June 10, 1935, Roy Poorman climbed out of the front seat of the OX5 powered Waco 10 biplane and said, "If you think you can turn this crate around, why don't you see if you can fly it by yourself; if you can't, at least don't scratch the airplane." Poorman taught over 200 pilots how to fly out of Wooster and Canton McKinley Airports between 1932 and 1963, many of whom became professional pilots. My own first license as an amateur pilot was C.A.A. non-commercial pilot's license #34876. Owing to my interest in becoming a geologist, I didn't get my private license #1,054,046 (SEL) until 1947 after my World War II duty as an AV(s) officer in the U.S. Navy's Bureau of Aeronautics.

Roy Poorman
1935

The time will come to write the story of my own flying career—not very dramatic, but for me, exciting and eventful, even if interrupted many times. Building time in flight was expensive, and except for refresher flight training as a veteran on the G.I. Bill at a cost of $300, all the flying I did was at my own expense. Flying as a student of geology opened the doors to both my career in the Navy's Bureau of Aeronautics and for post-war consulting work in exploration research to develop natural resources.

Few people have as raucous a sense of humor as do throttle jockeys who fly airplanes. Over the urinal in the men's room at the Coronado Airport north of Albuquerque, New Mexico, is a prominent sign that reads: "Will pilots with short pitot tubes and low manifold pressure please taxi closer." This is akin to the statement usually seen at airport restrooms: "We aim to please; you aim too please!" Some years ago, a waitress asked the late Henry Birdseye, a pilot who was heir to a frozen food fortune, "Would you like a glass of milk?" Henry answered, "No thanks. I'm flying today."

Roy Poorman knew Jimmy Doolittle long before he became famous as the pilot at the Cleveland air races; he flew the Gee Bee Speedster at 300 mph past the grandstand. In Jimmy's cross country flights for Shell Oil Company as head of the company's New York-based aviation department, he would stop at Canton's old McKinley Airport to shoot the breeze and tell jokes in many a hangar bull session. His zeal as a pilot was only slightly greater than the outrageous stories he told about himself and other pilots. Jimmy, when he was young, thin and all of 5'8" tall, once was cut off by a taxi cab driver as he was trying to park near Radio Center. Jimmy got out in moving traffic, tore open the driver's cab door, pulled him out of the cab, and cold cocked him on the spot. Everyone must know of Colonel Doolittle's famous raid over Tokyo early in World War II.

Flying an airplane is serious business, and many old pilots will tell you that "There are many old pilots and many bold pilots, but there are no old, bold pilots." To ground oneself after a flying career of some 40 years is a gut-wrenching experience. It has happened to many of us who, nonetheless, love to fly with anyone who will take us "up, up, and away."

The Pilot
1935

Part Three

THE ACADEMIC MAELSTROM

Last Day of School

I went to grade school in Berlin at the age of five in 1920. Our language at home was a *Niederdeutsch* version of German-Swiss with dialectic words of *Nederlandish Platte Deutsch.* For six years—through the sixth grade with Miss Edith Engel—the last day of school was very special indeed. A free summer ahead! All of us carried our lunches to school in lunch boxes, and on this last day each year, my mom would make a special picnic lunch. School was "let out," as the saying went, and then Miss Engel, a stentorian teacher who allowed no nonsense, suddenly became like a loving mother to us all. Under the pine trees on the north sloping lawn of old Berlin School, we would have a picnic. Our teacher brought special soft drinks, for we got thirsty playing the whole morning through; imagine recess from 8:30 to 11:30 in the morning instead of the usual 15 minutes during school time! Then the picnic!

Miss Engel on this one most-looked-forward-to-day of the school year brought a hand-cranked Victrola and 78 rpm records. She played record after record and sat leaning against a tree, dreaming of what we did not know. Miss Engel was a tall, handsome spinster who had a laugh that tinkled like a bell. Her voice had a golden tinge, but if students strayed in their attention, her voice was hard, cold steel. Many a time I had to sit on a dunce's stool in a corner in front of all six classes for some infraction which was fun and, I thought, harmless. Of course other students got caught too. But Miss Engel was never unkind, just tough! Only much later were we to find out why this lovely lady was a spinster.

On this last day of school was one outstanding event which took place each year. She played a final rewind of a World War I song entitled "There's a Long, Long Trail A-winding." It was a sad song, for she would tell us about World War I and the friend and a young suitor she had lost in France fighting the Germans.

On February 4 of 1978, I relearned the name of the composer of that song. He was a British war veteran named Vismic Capport who now lived in Venice, Florida. Much earlier, between 1947 and 1949 when I was first teaching at the University of New Mexico, I met a short, much older gentleman, a student taking special courses. He told me he had composed that song between battles in the trenches of France because he was homesick for England. We became friends when I told him about our last day of school in Ohio and how well I remembered how the song affected me. His name, Vic Capport, was never well-known in America. That meeting caused one circle of my life to be complete.

Knights of the Chalk Dust

Despite all the new modern visual aids available to the college teacher, the blackboard—now replaced in some classrooms by dark green hard-composition board— is still the basic surface combined with chalk and eraser, the tools to convey ideas. Large and small groups of students can be inspired by chalk drawings in white and various other colors. The process is simple, and variations in graphic ability are legendary. Professor Sam Knight of the University of Wyoming was a fine artist on the chalkboard; Professor Kirtley Mather scribed his lectures as detailed outlines for the Harvard undergraduates. Many other teachers made the blackboard literally sing with drawings illustrating new concepts. Obviously, one of the products of this frantic draftsmanship is chalk dust. It isn't unusual to see blue, red, green, and white chalk dust on coat sleeves of professors rushing from one classroom to the next. Since they are oblivious to this badge of their profession, the chalk dust is a bane to their wives trying to keep these absent-minded professors looking neat.

Woe to the teacher who misspells a word or gets lost in scribing a diagram. A hand shoots up with "Sir, what do you mean there?" At times a snicker runs through the class, especially if a mathematical equation is wrong. Professor Norbert Wiener, the great M.I.T. mathematician, would, it seemed, make errors deliberately in the vast network of equations he put on the blackboard. But one day he admitted that he simply couldn't do equations correctly on a blackboard! I know exactly how he felt. I remember that one of my mathematics professors in college—"Skippy" Knight, a wonderful teacher and patient savant—used an old high school system in college by having selected members of his algebra, trigonometry, geometry, and analytic geometry classes go to the board to "do figures," as he put it. He would read off one side of an equation and ask students to complete it. Then, because he had one short leg, he would stump from his desk and criticize our efforts. Very early, one becomes a sub-Knight of the chalk dust, at times wishing one could hide under the floor.

All the sophisticated new visual aids, such as taped demonstrations linked to motion pictures or serial color slides, will never replace the clear handiness of a chalkboard drawing, and nothing else will give the professor time to grope successfully for words of wisdom as he writes lists on a chalkboard. From the moment I first went to grade school, I fell in love with chalk—a marvelous invention! And to be picked by the teacher at the end of the day to beat the chalk out of erasers was a real thrill. However, cleaning the chalk tray was not that much fun, and my mother simply asked, "How can you possibly get so much chalk dust in your shirt and on your pants?" If the definition of a professor is "a person who thinks otherwise" or "a person who simply never got out of school," then certainly one of the major reasons for this is the love of chalkboards. Think of the vast array of knowledge that has passed from teacher to student throughout many decades of learning.

The University

The university, modern as it tries to be, is a bundle of contradictions, a conglomeration of poorly administered divisions, and, in a few ways, a throwback to ancient customs of the Middle Ages. It is an inefficient learning factory, dependent on outmoded concepts, lethargic in its teaching methods, scarcely a viable institution owing to its inefficiency. One does not open the head of a student and pour in knowledge. Universities today are perverse aberrations of the Middle Ages' situation where students went to study with famous savants and moved from savant to savant no matter where they were located. Students themselves demanded records of achievement, and thus was born the administrator to keep portable records. Eventually, instead of just knowledge, students insisted on grades which led to more records and more administrators to judge equivalencies of knowledge. Eventually administrators gained ascendancy over savants, grading them as to ability and rank, and concentrating them in one place or another for ease of control. The concept that has been lost is that a professor is the very highest rank that one can attain in a university. With the establishment of rigid standards, freedom of expression was dulled and knowledge was compartmentalized into a rigidity inimical to true learning. Administrators, formerly simple grade keepers, turned to housing teachers in classrooms and laboratories, limiting their movements. Learning was rigidified in schedules, the better to allocate knowledge in time segments. Soon it was found that students could more easily be controlled if they were also housed; it became a source of funds—never mind the supposedly poor food and crowded quarters. Today a student wishing to move from university to university to study with famous purveyors of knowledge is severely limited. This leads to competition among universities to hire the most notable teacher-researchers, overloading the numbers of ever-divisive departments and courses within departments and courses outside departments. The university then has to hire more and more people to teach fewer and fewer students about smaller and smaller segments of endeavor.

The great teachers are not the ones who do the most research or publish the most articles and scholarly tomes but are those who, as G.J.B. Smith put it, "have lighted the flame within us." In the labyrinth of academia, the only true way to become educated is to learn through one's own efforts. Teaching can only be a path to learning; the teacher can only inspire, not learn for the student. Inasmuch as commitment is the stuff character is made of, character in intellect lies solely within the dedicated learner. An educational think-tank in New York City, the Human Engineering Laboratory, has determined that the only acquirable characteristic for academic success is vocabulary and the use of words. That requires reading, aided by learning from a good teacher. All other characteristics of success in a person are inherent. Not all persons are created equal; only the opportunities to learn should be equal. Failing to light the spark may at times be due to faulty teaching, but, overwhelmingly, it is the inability or lack of desire in the student to learn. In this great country of ours, there is abundant opportunity for learning. Stupidity is inherent, but ignorance can be cured in anyone who will try. Colleges and universities that try to educate everyone equally know little but failure. There is nothing as dangerous as a highly ambitious person of limited intellectual ability. Not everyone should attend college or a university, unfair as that dictum may sound. But all should be allowed to try, and if they fail through lack of intellect, there should be no onus. Many people who work with their hands make far more money than do teachers. No test has yet been devised that will enable administrators to determine the difference with surety. Teachers have been surprised many times by students of apparently low ability who have been academically successful. I have been offended many times by controls that have stifled individual initiative, especially by university administrators and professors of mediocre capability who maintain that equality must be forced on the equal and unequal alike. Stated another way, there are vast differences in ability and ambition. "Mediocrity, wherever found, recognizes nothing higher than itself," says Howard Stein, a former manager of the Dreyfus Fund. Of all the places in the world of scientists, exaggerated egalitarianism in universities is debilitating for the capable who teach. John Wayne hit the nail squarely when he asked, "Is a fellow never again to get credit for putting in extra credit or for being smarter?"

The Harvard assessment of university administrators reminds one of what Dr. Carey Croneis, late Chancellor of Rice University, said: "A dean is a person who is too stupid to be a professor and too smart to be a university president." He maintained that good teachers were God's chosen people; with tongue in cheek, he continued by saying, "Teachers are [generally] people too lazy to work and too timid to steal." Those bits of jocular wisdom came from one of the great geologists who was a famous professor, author, chairman of a department, and later chancellor of a major university. And finally, Dr. S. J. Shand, that brilliant petrologist from South Africa who became a distinguished professor of geology at Columbia University, philosophized that "A university is not a lecture theater or a library or a laboratory or a place at all; its essence is a frame of mind!" Where two or three are gathered together in the name of knowledge, there is a university!

My office at the
University of New Mexico

Scientific Courage and the Challenge to Change

As a member of several executive committees of national scientific organizations, I came to realize that exploration scientists have broad capabilities—beyond basic science—and applications to find and develop energy resources. During visits to scientific and professional meetings, as a participant on geological field trips, or after attending affiliated society meetings and annual conventions, one can only be astounded at the executive capabilities of geologists who administer the various activities of such large organizations and their affiliates.

It was my privilege as a major company geologist, as an engineering Naval officer during World War II, and as a professor at a large state university for 30 years, to work with thousands of students, many hundreds of whom graduated to positions in exploration on federal and state surveys and in teaching. To all, while they were geology students, I insisted that they do the following: 1) become members of local geological societies; 2) volunteer to serve on committees, give lectures, lead field trips, publish papers, edit guidebooks, and serve as officers (in short, as soon as possible, use their spare time to learn the difficult art of becoming executives, not just run-of-the-mill job-holders in geology); 3) give of themselves through volunteer work for organizations, exceeding, but not neglecting, their regular exploration work as geologists; 4) above and beyond all that, be ethical in all their activities, have courage as scientific professionals to work beyond what is expected of them but also expect proper compensation, be unafraid of their superiors, and be ready to learn from their underlings, peers and mentors within the inevitable complexities of all organizations.

As scientists we are in debt to those who preceded us; certainly, we are in debt to all who were our mentors whether at college, the university, a job, or a volunteer organization. Yet we should recognize that learning must continue, adaptation to change must go on, and no matter what we may have contributed to the geologic literature ourselves, we must read, read, read, and think! Many geologists, who merely hold jobs rather than pursue goals, pride themselves on being conformists instead of exercising their capabilities to seek solutions to difficult scientific riddles. Why not rise to challenges even to the point of nonconformity? Rebuffs are a normal part of the life of any geologist whose success leads to jealousy on the part of the less capable. Luckily, we are not all created equal, yet anyone who disdains those who are less capable deserves to fail in human relations. Shoot for the moon and let the chips fall where they may!

It has always struck me how adaptive exploration geologists really are. Given a chance, a challenge or an apparently unsolvable problem, the fervent geologist rises to, and beyond, the situation and so becomes a beacon for those who follow!

The Professor

There are as many differences among professors as there are professors, for one thing is certain: a professor is a person who thinks otherwise. Some are practical scholars, others aren't scholars of any kind; some are so impractical as to require careful nurturing so they can function in society—that is usually the role of a professor's wife. Many are so boring that they illustrate Sir Peter Scott's dictum: "A lecturer is a person who speaks while others sleep." Another type is the scientist who knows everything and is always right. These are the dullest of all individuals. When a man displays his erudition too readily, be watchful, but if he is a professor, he's being paid to be erudite. A wise professor—and such do exist—will follow the advice of Kung Fu: "I seek not to know the answers but to know the questions." When I found that quote 17 years ago, a notion I thought quite bright struck me: Why not give an examination on which I will list 20 answers in detail and then ask graduate students in geology to compose cogent questions in response? The first time I tried this method and saw their questions, I realized that I had found a technique to separate the thinkers from those less able to think—a valuable tool which began to direct the type and range of teaching necessary for an advanced class in geology. Arthur Laffer, that tongue-in-cheek author of the Laffer Economic Curve so favored by President Reagan, had this to say about professors: "You should never let a professor run your country, but you should listen to what he has to say." Some believe a professor is really only a super-annuated student who never got out of the university.

Being a professor is one of the few professions where one can assume center stage without rehearsal, but where it is needed most. To move directly from one's own graduate work to being a college professor almost insures failure as a good teacher. Today in science, a post-doctoral position in a university or research institute leavens the individual, gets him out of the classroom where he only received information, and puts him into a non-teaching position where he must produce obvious results. Such a situation was not available to me many years ago. I recognized that if I was to become a successful teacher, I had to go into industry and become a practical scholar. To teach only what I had learned at the university—even though attending scientific meetings, doing my own research, and keeping up with new scientific literature were diligently pursued—was NOT adequate. One major effect of research in industry is that one's theories get tested very quickly and thoroughly. Failures will occur, but one learns to savor the hunt and soon develops the guts to look failure in the face and try again and again. There are important lessons in failures as well as many satisfactions in being proven right. One old saw in petroleum exploration is this: when a dry hole is drilled, "it's a geologic success and an economic failure." Where dollars are concerned, a well may be a failure if there wasn't any oil there, but facts involving theories may be proven. Yet that failure is a stepping stone to another try and possible success. In all of the sciences, such testing is absolutely necessary for a graduate who hopes to become a professor.

Of course, becoming a geologist in a large corporation has certain drawbacks. One becomes a small cog in a wheel among many larger wheels. As to the cog difficulty, I'm reminded of a humorous statement made in 1978 by an Exxon geologist named Don Young: "Exxon's like a log covered with ants moving down a river; every ant is convinced that he is steering the log." A major natural resource corporation, nonetheless, is always years ahead of a university in the broad field of exploration and production concepts in geology. A corporation is filled with successful scientists and engineers. Training with them is far beyond the education one receives in a university. Also, an economic impetus causes the ivory-tower theoretician to become a practical scholar—of whom there are far too few in modern universities where practical research is thwarted by governmental research grants.

Geology is a tight combination of all the sciences, yet it is also an art. Geologists so often must make major decisions with few facts; this may be an unusual indicator of how a professor should become proficient as a teacher. One of my students, the late Grant Wilson, a successful geologic consultant, once said, "Geology is not a profession; it's a disease like malaria." Using logic a teacher of geology may be teaching a disease that can't be cured. This seems to be true, for "once a geologist, always a geologist," and few of them ever retire completely. A student of mine, who longed to become a professor, took a teaching position at a university as an instructor after getting his master's degree. In a year he was back to study for a doctorate with this gem of wisdom: "A university is the location of the highest level of covert back-stabbing in the whole world."

What most administrators and few staff realize is that a professorship is the very highest position in a university and thus a heavy responsibility, for teaching young minds to think is a high calling and an activity of some splendor. It's too bad many professors drudge along year after year, teaching the same old stuff from old yellowed notes from courses taken in graduate school! Ben Sira, presumed by some scholars to be the author of Ecclesiastes, said, "Let us now praise famous men." He went on to single out students of excellent teachers for special approbation thus: "He will grow upright in purpose and learning, he will display the instruction he has received...The assembly will celebrate his praises." There can be no finer encomium for a teacher than to hear students so honored.

With students in Tijeras Canyon
1981

The Liberal Individualist

A professor teaching at a university confronts all types of students, ranging from the meek to the rash, the stupid to the intelligent, from the conservative to the ultra liberal and the poor to the rich. The polyglot American is a rich mixture of political thought; all are tremendously interesting though not all are interested. Of all the 17 different courses I taught in my 30 year career, some six were seminars in fields in work I was qualified to teach through experience but fields in which our university had no formal courses. Of the 11 others, I instituted eight of them to enable senior, masters or doctoral candidates to complete programs of education to meet the needs of the burgeoning science of exploration geology. The estimate of some 6500 students who were in my classes may be in error—I still have all my grade records, and one day I threaten actually to count the exact number. For now that is not as important as it is to note that they represented a remarkable cross section of America's hearty interest in politics. "There's a diversity in discontent in the hearts of Americans," said radio commentator Paul Harvey.

One does not teach entirely objectively in geology, for that eclectic science is also an art, the art of discovering the possible truth about the many vagaries of the earth's history. Every student soon learns that it is a retrodictive science predictive of the location of valuable resources. As such, we more than touch on economics, politics, sociology, psychology, archaeology, anthropology, law, business, plus administration. Geology is a science based in the description of the earth and the appraisal of the origin of its features in our attempts to predict what will happen next.

In the roaring 1960s of student activism, students began to challenge professors and the status quo which some thought was the basis of our teaching. At the same time, a revolution was occurring in our knowledge of shifting continents, growing oceans (i.e., the Atlantic), and closing oceans (i.e., the Pacific). It was difficult to prove to students that a professor of geology was also a serious student who tried to explain the earth, the function of man as the seeker, and the effect of political action on exploration for and utilization of natural resources. Students rarely could understand that not only were all natural mineral resources finite, but that renewable resources were also dependent on rocks (from which the remarkably thin soils are made to give us ALL our food) and a great symbiosis of rock and soil to the genesis and maintenance of our atmosphere, hence our weather and climate.

Once the student grasped the concept of how basic and slender is the thread upon which life on this planet depends, it was but a short step to a brilliant recognition—almost a flash—of how dependent the races of man are upon each other. There was born in some the politics of survival, the careful use of resources, and, thus, the functions of conservation. Many times students would come up after class and ask, "What is your political affiliation?" To answer that took some thought and made me wonder, "How could such a personal question come out of a class in geology?" These students were concerned about the earth. They were taking courses in fields which I once recommended should have been lumped into the one department of socialism. It became a battle to teach them just how dependent the human race really is on rocks, air, water, and sunshine—rather than on strangling political systems, full economic control by governments run by imperfect clerical-type politicians, confiscatory taxation that stifles resource development, and tight control on political thought. Thomas Jefferson said, "Government is a singularly dangerous beast and must be chained."

The expected answer was, of course, Democrat, Republican, Independent, Socialist, Communist, or else Libertarian. My answer always was Liberal Individualist. They immediately detected the irony of that answer, the very ambiguity of such a concept. "Well, how do you vote?" In a nice way I would say, "It's really none of your business, but as a registered Democrat (at that time) I always vote for the best man or woman running for office! Will you be able to grasp this? I have never voted for a Democrat for president." Actually, when we moved to New Mexico in 1947 one simply could not vote unless one was a Democrat. My father in Ohio was a Jeffersonian Democrat, violently opposed to Franklin Roosevelt and Harry Truman. By 1980 I changed my voter registration to Republican, for that is really what a Jeffersonian Democrat is!

Finally in these political discussions with students, I asserted my right to be a rugged liberal individualist who asked nothing of government but to be left alone and to be liberal in thought so that any human living under our form of government has the right to do whatever he pleases, and is capable of doing, without impinging on the right of others. Howard Ruff maintains, "The Constitution is not so much an exercise in government as an exercise in protecting us from government." This is recognizable as the ancient Mennonite code. James Madison said, "If men were angels, no government would be necessary." But the thrust of teaching geology was, as a liberal individualist, to stress the importance of the statement by Henry J. Curtis: "There is nothing as frightful as ignorance in action."

Geology Building
College of Wooster
1935

Professor of Geology
Karl Ver Steeg
College of Wooster

History, Anyone?

Thomas Carlyle believed that "Biography is the only true history." If such be true, the study of history could be made much more interesting if it were all cast into the form of biographies of men and women. Rather than study governments, annihilations of peoples and races, great migrations of displaced populations, and dry renderings of dates and statistics, why not study the developments and doings of people who actually made history? Most historians would laugh at such an approach.

The study of history, in whatever form it is cast, is highly necessary. Someone once mentioned that people who don't study history are bound to make mistakes previously made. When Ben Franklin said, "Where liberty dwells there is my country," he was suggesting that the fight for liberty is an ongoing activity, and it only can be maintained by vigilance plus a study of the loss of liberty in times past.

When I was at a liberal arts college studying geology, students were required to take courses in psychology, philosophy, foreign languages, English, history, and other sciences in support of the bachelor's degree in geology. Economics was recommended but was a washout as far as learning about money. History was taught by two professors: one covered ancient history, the professor himself rather ancient; the second, a young man, covered modern history. The general break between the two was the birth of Christ, logical in as much as the college was of Presbyterian persuasion, though no longer financed by the church.

Today, 60 years later, the same college has a massive

department of history with eleven professors and full-time instructors. When I was in college during the early part of the Depression, there were 900 full-time students, 450 per history professor. Today there 2200 students, or 200 students per history professor. History, taught as a social science, has been segmented into many sub-divisions, all seemingly bent on creating a state of socialism in America. Predicated on the concept that history with its lessons is perceived as the perfidy of man, or humans' penchant for taking advantage of other humans, history now is dedicated to the activist approach of creating an ideal future in which government will take care of everybody to the detriment of individualism.

Within the same span of time years ago, top professors initially taught four courses per semester and were paid $1800 to $2800 per year. Today instructors are paid up to $16,000 per year for teaching two or three courses, and the highest ranked professors are paid $36,000 to $60,000 per year and teach only one or two courses per semester. Is it any wonder that tuitions which were $150 per semester in 1932 are now over $5,000 per semester in many liberal arts colleges? Major universities have also skyrocketed total costs to as much as $20,000 per year. Elitism is to be commended if all able students could participate with opportunities to match their abilities. Where today can an underfinanced student get a formal education? Forced to live at home, the answer seems to be local junior colleges, tech schools, and state universities—and they can be sources of great educations under splendid teachers.

The teaching of history, itself an esoteric subject, is emotionally desirable but is hardly a preparation for a career. It mirrors the changes in almost all departments of colleges and universities. Could one go back to two professors of history in a college wherein history is a required subject needed to broaden a student's horizons? Not hardly! I dare the reader to name one person, other than a modern day Will Durant, who could be found able to teach either ancient or modern history as the inclusive, generalized courses so necessary for a well-rounded education. What has happened to those inspiring teachers who challenge us to continue the study of history, one of the most fascinating studies within human ken? Instead, the modern historian is bent on inculcating a narrow discipline rather than inspiring students to the broad philosophy of our historical background. They are now businesses packaging a product rather than truly educating our youth!

Campus buildings,
College of Wooster

Professor Sherman A. Wengerd

Economists—Are They Necessary?

It is a rather well-known fact that economists are usually ineffectual in the hurly-burly of the business world. It escapes me, then, why government persists on hiring economists rather than successful businessmen as economic advisors. One answer lies in the experience of Robert Packard of Hewlett-Packard, an eminently successful businessman twice called into government service only to leave Foggy Bottom crestfallen that he couldn't defeat the lethargy at the Department of Commerce which he was hired to correct in Washington, D.C.

Few economists become businessmen because it's easier to talk about money than it is to work hard to make it. That's why so many economists become professors and, perhaps later, bureaucrats, talk being easier than action. To feed at the public trough is so easy compared to risking one's neck in making, buying and selling products and dealing with labor. Perhaps it may be noted that few bankers are good economists, if such truly exist; in fact, as risk-taking businessmen and investors, bankers are real duds. Of course they are dignified, else why would we give them our money at five percent, so they can loan it out at ten or more percent while they lose money on bad loans, overstaffing and fancy offices in new buildings? Thomas Carlisle called economics the dismal science.

The center of focus in banking, economics and business in our capitalistic system is money. Ben Franklin said, "Money, more or less, is always welcome." His Poor Richard averred, "A penny saved is a penny earned." But Guy de Rothschild, who had more than a little to do with money, thought, "Money is the scapegoat for our misfortunes." Who has not heard of a man or a woman who spends money like it's going out of style? Pik W. Botha, finance minister, who, when asked if sanctions would cause shortages and bring South Africa to its knees concerning apartheid, stated, "If one has got money, one can buy anything." H.L. Mencken added a touch of humor by noting, "Wealth: any income that is at least one hundred dollars more a year than the income of one's wife's sister's husband." Economists should recognize in Mencken's statement a basic truth: wealth is relative, no pun intended.

My friends in the economics department of 20 years ago at the university were all socialist-liberals who believed that government is far wiser in handling economic problems than are individuals—whence has come the socialistic concept that the whole economy should be government controlled. "The people don't know what's good for them!" This kind of thinking has been in vogue from 1932 to the second election of Ronald Reagan. The government of the United States is still the repository of more economists—who can't make a living otherwise—than all the universities put together. No wonder we're the greatest debtor nation in the world today.

I studied economics in college in 1934 with Dr. Alvin Toslebe, a Keynesian and Rooseveltian New Dealer. Why he left the government for a college teaching job that paid only $1800 per year has never been clear. In his classes socialism was preached, individualism was to be shunned, all money was taught to be handled by government, taxation was instructed to be very heavy, and people were to be cared for from cradle to grave. In one of his particularly virulent lectures during the class year of 1934-35, he asked the class if there were any questions. I stood up and asked for a vote of the class to see how many believed what he said. Before anyone could vote, he dismissed the class and rapidly disappeared into his office, slamming the door. Of course I got a C in the class despite high averages on all the tests. Thereupon I refused to take any more courses in economics—one of my favorite subjects at the time of severe depression in the U.S.A., 1932 to 1936. Why take economics in college? It was recommended by my professor of geology because my major field of natural resource exploration required a solid base of economic theory. Reporting this class to my geology professor, a rugged individualist in the ken of my businessman father, he stopped all recommendations involving taking economics.

Like cries in the wilderness, men and women are rising to teach sound econometric courses, and such thinking is now being taught in some universities as business administration courses. I have recommended to our university administration a number of times that they merge the departments of social science, economics and government into one department of socialism and transfer at least half of the history department into this new department. This idea, they say, won't fly. Fortunately our College of Business Administration is now the bellwether of logical economic thought and teaching and is growing, while the departments above named are withering on the vine.

There isn't one solitary humorous thing I can say about economists. Many have tried. It's best to say, however, that one of my best friends is a retired professor of economics. I'm happy he's retired!

Finally here is a letter by Leonard Silk—published first in the *New York Times* and republished in the January 1, 1976, issue of *Forbes*—concerning one man's view of economists. It was titled "Economics Is A Profession?": He said, "The economics profession is caught in a curious and complicated paradox: The worse economic policy is, the worse the state of the economy; but the worse the economy, the more people turn to economists to solve unsolved problems, the greater the sense of dissatisfaction and inadequacy within the economics."

Dad and Mom Wengerd with my sister, Carol Ann
April 1939

The Pecking Order

Capt. Bruce Barnes, former owner of the *Motor Vessel 105* out of San Diego—reviled by First Mate John on a voyage to Scammons Lagoon for leaving dirty knives on the snack tray in the lounge—retorted, "I can leave all the knives dirty I want to. I'm the captain!" The captain then blasted forth with an old British Navy saying of the 17th century: "God sends meat, but the devil sends cooks!" Because of mutinous leanings by sailors poorly fed, the normal pecking order at sea is greatly resisted by cooks. Like cowboy cooks on the open range on a cattle drive, the cook is a supreme boss within 50 feet of the chuck box.

Perhaps in no other part of society is the pecking order so closely guarded by all participants as among sailors on a ship or workmen in work gangs. In days long gone, the man who could whip everyone became the leader and dictated the pecking order based on whomever he liked or disliked. In the culture of India, such order is solidified into castes almost unbreakable by individuals low in the order of things. In the U.S. Navy, rates and ranks are carefully adhered to, but superior ability and the desire to rise via education allow for the penetration of artificial ceilings. The pecking order in universities has dominant control among staff on wages, but teachers on contractual salaries can move through ranks from instructor to full professor through valuable service and time. Such a rise based on achievements in teaching and research—under the watchful eyes of peers and mentors—are often thwarted by administrators.

One need only run a chicken farm to realize that the

biggest, fightingest rooster is the cock of the walk. In ancient societies and, indeed, in many third world countries today, the man with money, or the man with the most guns, soon becomes the richest man. These leaders usually reign by force. What about democracies? Here the rise of leaders is based on popularity, and that is all too often based on money used to convince through advertising that an individual and his retinue can give the greatest benefits to the greatest number of people. More directly, votes are bought, which means the pecking order based on ability is fractured by money and influence.

How does one keep from getting trapped in an unfair, or any other, pecking order? How does one fight the inequities? Any country boy from a small town, avoiding the sexual snares of an early marriage, alert to the dangers of a wage-earning job, keenly observant of the rich and famous, unafraid of hard manual labor (but hating it), having average intelligence, and undeterred by long hours of study can move through the ranks of whatever line of endeavor he chooses. Of course it helps to go through an economic depression between the ages of 15 and 22—that is, being unable to get a salaried position of one's choosing. That circumstance keeps a person in school as long as tuition can be earned or won through scholarships and fellowships. Stay at it long enough with enough vigor, and one can ride a university career to the end of one's days. One must be smart enough to avoid becoming a university administrator though; here, the greatest pecking order syndrome is suffered in the academic society, and back stabbing is a fine art. What's the answer? Open with your strongest suit, and choose your pecking order with care!

The Committee

There are few tasks more onerous at a university than serving on a committee. Yet some professors serve on many committees out of choice; seemingly, they have nothing better to do. A quick survey of such people has led to the conclusion that the social scientists—who aren't really scientists at all—lead the pack in attending committee meetings. They find such meetings interesting and are usually the most voluble activists. It seems a small person must belong to something bigger than he is, but a great person must beware of tyranny by the common herd. I once served on a committee with a scientist who told me in confidence that he brought only inconsequential matters to his committee; the big issues he decided himself and so reported his decisions as committee decisions to his dean. As an ad hoc departmental chairman during several summer sessions, I noted that the wrangling that goes on in a select committee, known by the grandiloquent name Committee of Departmental Chairmen, was amazing. Later the term "chairperson" came on the scene thanks to the ultra-feminists and lack of backbone in administrative men. A true travesty in our language!

There are many definitions of a committee, and you may have heard them all. However trite, all are true! Take the one that says "A committee is a group of people chosen by the unknowing to decide things that are unnecessary"; or, "A camel is a horse put together by a committee"; or, "A committee is a group of the unfit appointed by the unable to decide matters that are unimportant." The one I like best was by that peripatetic sayer of wise things, Anon: "We, the uncomplaining, using the non-existent, willingly do the impossible for the ungrateful."

Mediocrity glows in a committee; in fact, it is in a committee that a mediocre person can gain some measure of fame by riding on the mental coat tails of more than one able person among the really capable faculty members who make up most committees, regardless of size. Spread the honors of a good decision, that is, one that the university administration likes. However, a committee is a bulwark against righteous indignation if a bad decision is reached. I once served on a tenure committee in which the chairman was instructed by the academic vice president to deny tenure to a nationally-known scientist because he married the unhappy wife of an especially prominent businessman, a pillar of the community. Talk about a real brouhaha in a committee. The chairman, of course, told us what he was told were all the bad things about the scientist, but several of us made investigations on our own—now, there's a real no-no for committee members. When the vote came to grant tenure, because the fame of the man was international and his teaching and publication records were superb, the chairman of the committee quit in a huff and the academic vice president was livid. The president approved the committee decision, and academic freedom was upheld. I made an analysis of just how famous the academic vice president really was as a social scientist and found that his so-called international reputation extended all the way from the administration building to the VERY EDGE of the campus. So much for fame and administrators.

Generally speaking, a man who walks in fear walks nowhere. In a research committee to which I was appointed by a dean, I announced to the committee chairman that it was the dean's duty to allocate research funds—meager as they always were—and not that of a committee to vote on funds for their peers. In a matter of weeks, I was transferred to the University Athletic Committee, a real hotbed of revolution, for the faculty was adamant that we do something about the high salaries of coaches. There is no situation that red flags an administrator more directly than high salaries for coaches as compared to low salaries for faculty. Our decision acclaimed by the president who had himself been a well-known athlete was: Coaches should be paid more. They're more easily fired than a tenured faculty member. Social scientists rose up in arms in the faculty senate and demanded that the Athletic Committee be fired forthwith. It seemed strange to note that the next year the firebrands of the faculty senate had left because the president simply didn't give them raises—truly a momentous non-committee decision.

The academic world is a wonderful place; only people can louse it up and generally they are committee members! I've tried to analyze what makes people enjoy committee work. This has involved committees of local, regional and national geologic societies, institutes, associations, and universities, including women's committees (which are more like social clubs). Committee work expends time of people who do not have to be busy doing useful things. Meetings are times to relax—unless the subject involves one's own ox being gored—a time to sit back, smoke a pipe, look wise, and sleep while others talk. I know one committee chairman who had his committee meet in a very cold room. You can't imagine how fast that committee made decisions. In another committee, the chairman wore shorts on a very cold day, turned up the heat in the room, so everyone went to sleep right after lunch. When he called for a vote, it was sleepily unanimous, no one remembering what the vote was about.

All in all, the committee system seems to fulfill a need among humans to try to decide things. Spreading the responsibilities among committee members is the job of the chairman. Using *Robert's Rules of Order*, such meetings can be efficient and do valuable work for departmental chairmen, academic deans and vice presidents, and presidents of colleges and universities. But does everyone have to serve on a committee? Do members serve to protect their own interests? Are feelings assuaged that the committee system protects faculty from dishonest moves by administrators, or is it a jolly way to spend time visiting with peers? Of long time committee chairmen, one can say, as William James did, "No priesthood ever initiates its own reforms."

Teaching, Tenure and Obsolescence

The very easiest place in the world to become comfortably obsolete and still earn a living is at an American university. Dr. Odie B. Faulk, a professor of history at Oklahoma State University, said in 1973, in reference to teaching, "This beats working for a living." When I was teaching historical geology at the university, I discoursed on the immortality of single-celled life, which subdivides, or buds, into numerous individuals that do the same, ad infinitum. I said, "It's even possible there are single-celled creatures on this campus, a part of the original cells of Precambrian age. I might suggest to you students that there are some faculty members who may remind you of that aged condition." That statement got around to several departments that had ancient professors, long obsolescent, who should have retired to make room for more capable younger teachers.

A graduate dean once got up in a faculty meeting back in 1950 and stated, "There are too many tenured full professors and tenured associate professors at this university; we cannot allow this ratio to increase, so we are not going to advance assistant professors for several years." Many of us assistant professors saw our progress in danger of halting. A few who had already become nationally-known left the university. I got up in a faculty meeting (I was not yet on tenure) and said, "It's obvious. If the ratio is wrong, retire some of the unproductive old timers and hire more assistant professors!" This riled up a few administrators, who also should have quit years ago. Fortunately, I had three offers for higher ranked and greater salaried positions in three more prestigious universities by 1951, which I duly reported to my new dean. During the fourth year, I was assured that I would be raised to associate rank in my fifth year.

The old order changed, proving that a national reputation earned early is the best defense against a stupid concept of limiting ranks by administrators. At the end of my fifth year, a new academic vice president position was inaugurated. A kindly older professor of ancient history called me in to his office and said, "You'll be up for tenure next year." I told him I didn't want tenure, that I'd rather have yearly contracts without tenure. He was shocked, and then he said, "Dr. Wengerd, if you don't accept tenure, we'll fire you"—a shocking statement from a nationally-known expert in Spanish history! So I accepted tenure because I wanted to stay at New Mexico for a number of excellent non-academic reasons.

In 1944 during World War II, between sea duties, I would at times have breakfast with an old friend, Dr. Claude Needham, in the Interior Department cafeteria. Dr. Needham had been president of the then New Mexico School of Mines, and I had met him on a field trip in 1941 in New Mexico while I was a geologist for Shell Oil Company in Amarillo. I asked him about teaching at the University of New Mexico. Would such be a good move after I finished my doctorate at Harvard, when and if World War II should ever end? He made a statement which I would never forget: "Yes, it's a good school, but you will not wish to stay." I did not ask him why.

It is sometimes forgotten, but a well-known fact is that you cannot teach what you yourself have not done. After two professional careers—one in the Navy, the other in a major oil company—I knew I would never be afraid of losing a job in teaching or anywhere else. I learned early that George Bernard Shaw was right when he said, "You don't learn to hold your own in the world by standing guard but by attacking and getting well-hammered yourself." If you want to become an administrator rapidly at the university level, move wherever you can gain half again as much salary and a higher position. Taking a good teacher who enjoys teaching out of the classroom to make him an administrator is academic malpractice, yet many professors make that move in order to gain higher salaries. Twice in my teaching career at the University of New Mexico I was almost forced to become chairman of our department—an onerous clerical job at worst, a position of doubtful two-bit power at best. Both times I said I'd move to another university, and the administrator relented. I simply told him I already held the highest position at the university, that of a teaching research professor. Albert Camus suggested that the first thing is to learn to rule over oneself. Why become an administrator when it's tough enough just to administer oneself? Professor Peter was right: people do make the mistake of rising a level or two above their competency. It happens in every university!

In my 30-year teaching career, I taught 17 different courses, and as our staff was increased, I arranged to teach just 10 different courses in a two-year cycle. In my first 10 years of teaching four courses each semester, the academic revolution took over, and I was ordered to teach no more than three per semester. At retirement I was only teaching two courses a semester, though in my field three were necessary. That should tell the reader something about academic padding!

Stanley Kramer in the *Seattle Times* had this to say about teachers: "A mediocre teacher tells, a good teacher explains, a superior teacher demonstrates, a great teacher inspires." All four levels of teaching exist in a top rank professor of some experience. A person of acute sensitivity to the type of students in his classes uses all four techniques, and most of us early in our careers tell, advance to explanation, thence to demonstration, and finally to inspiration. Even in a single class, this may be a progression during the teaching of any group of students in one semester. In some complex subjects, one may tell, then explain and finally demonstrate, while inspiring the student to think ahead from a base of what transpired earlier in the class.

A friend of mine who had a Ph.D. in French but became chairman of our journalism department—the late Dr. G. Ward Fenley—was intrigued with a statement I made to him early in my teaching career: "I can hardly wait to get to my classes to tell them what I know, and how to go about learning more and thinking for themselves." He mentioned that such an attitude was the very guts of good teaching. It takes a large ego to teach well, plus a reasonable amount of theatrics. In some cases with complex concepts, I warned students that I would exaggerate, even at times tell them sequences or facts that were not exactly true. Shocking! But always, I would warn, learn it this way now, then as we go along, we'll modify, test, experiment, and you'll learn more definitive and correct information.

Bob Grant, a student of mine, a well-known consultant who became a representative in our state government, remembered well my admonition to "Learn it my way, then don't let what you learned stand in the way of thinking far beyond what I taught you and refute what was wrong. Let the chips fall where they may." Also, teachers make many honest errors. Andy Capp in 1974 said, "If there's one thing worse than being in the wrong, it's being in the right with nobody listening."

My very worst case of a misguided professor involved a man who was a consultant to the detriment of his teaching career. He told me in amazing candor that "I never teach students all I know because they might become competitors of mine in the field of exploration." If I were more specific, many geologists would immediately know his name! Such an attitude is academic dishonesty. The whole thrust of teaching is to prepare students to become far more famous and successful than the teacher. Their elevated positions then rebound favorably to the reputation of the teacher. Numerous past students of mine have come back to tell me how much greater their salaries were than mine as a teacher. I congratulated them, and they became non-plussed when I told them, "That makes me a true success!"

Senator Alan Simpson of Wyoming struck another note when he said, "Professors sometimes blur the line between divinity and tenure." To attain tenure is indeed considered to be divine and to be a defense against rapacious administrations with whom a professor may violently disagree. The so-called Freedom of Speech argument infers that professors not yet on tenure are liable to speak carefully whereas a tenured professor can say anything and is always right.

Today it is not unusual for tenure denial to hit the civil courts. More and more, teachers with doctorates are taking positions outside the tenure track by accepting one year contracts. There must, at some point, be honest controls on who should have tenure. But the great problem in universities now is to get rid of an obsolescent teacher who has lost his capacity to inspire students. Today, the strong suggestion is to leave at age 70. The more intelligent ones retire at age 65. Some of us who have other fish to fry, far from obsolescence, retire at 60, 61 and 62. By such self-imposed early retirement, all universities would be strengthened with vibrant new teachers "anxious to tell students all they know!"

Taking to the air with students.
I am in the center.

Time to Retire

When even the young new faculty on a campus begin to look like students, it's about time for us old-timers to retire from teaching. As a long-time professor in a state university, it was not unusual to have students come up after class and say, "My dad (or mom) took one of your classes." That gave me a good feeling that one is remembered after such a span of time. But in 1975, a bright cheery girl came by my office to say how much she enjoyed historical geology and, incidentally, said she, "My grandfather took two classes with you." I went into shock, and the next year I retired. Her grandfather had been an older veteran of World War II with grown children; her mother also went to the university after an early marriage,so here stood this chic youngster, a granddaughter. I told my wife about that, and she said, "I wanted you to retire earlier."

As a young geologist, I was fortunate enough to be elected national editor of the Bulletin of the American Association of Petroleum Geologists. After what seemed to be only a few years, I was invited to join the "Society of the Retired" of that association and pressed into service as Master of Ceremonies at their annual meeting—"Round-up of Old-Timers"—held at the Cowboy Hall of Fame in Oklahoma City. How can this be, I wondered. Where did the years go?

To a person nearing retirement from an eight-to-five, five-day-a-week job, retirement holds unknown terrors. To have her husband underfoot at home, led Mrs. Casey Stengel to say on Casey's forthcoming retirement, "I love Casey but not at home for lunch." But a scientist, especially a geologist, seldom retires. He or she may quit teaching but never ceases to look at rocks and countryside to see how it got that way. Some professors leave the classroom to do consulting work at their own leisurely speed; others write books or write letters to the newspapers, or travel, or go into investments where a geological background is a splendid aid to investing in natural resources. I know many geologists who became wealthy by developing oil and gas royalties or production after long teaching careers. Some of those began to plan just that and laid the ground work while still teaching. Their retirements were smooth transitions to interesting new careers. One, Andrew Lawson of the University of California, married his young secretary, started to build houses, and made so much money in construction that he was heard to say, "How long has this kind of money-making been going on?"

A former boss of mine in the Shell Oil Company, where he was exploration manager, resigned to become vice president of a medium-sized, old and very successful oil company in Tulsa. In Shell, buying royalties was forbidden, but in his new position, he was allowed to make royalty purchases. Within only a few years, he became a millionaire—all ethical and legal and with the blessings of higher management, meaning the president, who was doing the same. One day my boss at the age of 75, having done consulting work after retiring from his vice presidency, took his maps and cross sections out into his back lot and burned them all. His next career for almost 20 years was to make investments and plant a large garden. He did this every year until he was 94 years old. Retirement was obviously no shock to him; he'd done it several times and was successful in everything he did!

Trever Newell said, "Being retired takes all the fun out of Saturdays, but it sure puts it back into Mondays." If one looks at retirement as failure, it's bound to be a failure if one doesn't have the guts to try new things or put new twists on old things. Henry Ford II said, "Don't complain, never explain." Retirement requires changing gears. Realizing that time is short allows one to say "take your time but get crackin." And complaining about not having anything to do is rampant in the land where so many people are retired. Winston Churchill, tossed out of government by the Brits who reviled his direct approach to solving problems, took up in earnest an old hobby—painting—and wrote a remarkable set of books involving World War I. Besides that, he never gave up good cigars and brandy. It's well to remember, however, that it's not wise to stay around until one becomes a legend!

The right to fail in retirement presupposes the right to be successful, and both are based on individual freedom of choice. The earlier one makes the decision to save half of all the raises he ever earns and at least 10 percent of his salary, the wealthier a man becomes. This truism was taught me by my first boss in the oil business. Anyone who thinks Social Security payments after age 62 are sufficient for retirement is bound to go into severe shock. Even strenuous saving is not enough, thanks to the pernicious taxes on income, property and purchases, plus inflation, which is a normal phenomenon in a democracy built on money. If by the age of 35, a bread winner hasn't set up a retirement plan and stayed with it through thick and thin, he's bound to be one of the many of our population who haven't enough to live comfortably after retirement. That retirement plan must include wise investments in which money earns money, or purchase of real estate earns money, or development of natural resources, both renewable and non-renewable, earns money. Poverty is no disgrace, but it is frightfully unhandy! The man who expects to get ahead by working only eight hours a day will do just that: expect.

There is no one right way to retire for everyone. Given better diets and greater freedom of choice, many more will retire in the future than have retired in the past. We all wish to make our own decisions about all ranges of problems, from buying a pocket knife to what to do about dictators. But if our desires are satisfied by the decisions we make in our own views of the facts, we are successful, whether anyone agrees with us or not. Ralph Waldo Emerson, intellectual savant, went to the Cambridge jail to visit Henry David Thoreau, who was arrested for some supposed misdemeanor against the local bureaucracy—probably for not paying onerous taxes. When Emerson asked, "Henry, what are you doing in here?," Henry replied, "If you were wise, you'd be in here with me." That interchange—not the exact wording—was relayed to me in great glee by one of Emerson's descendants. Thoreau saw life from the view of a different drummer. Though he was a Harvard graduate, he chose to do early in his life what many of us look forward to in retirement: contemplate and smell the flowers.

The prime principle of retirement is don't let anybody stampede you into doing any work. It's rather like coming back to the freedom of the summers between grade school, to have no more classes to teach at a university nor the monotonous regularity of meeting eight o'clock classes, even though teaching was great fun. As soon as people know you're about to retire, the offers to do work for money triple, and dozens of people will say, "You're retired now. What are you doing with your time?" Depending on how well I know the questioner, my retort is "I'm loafing" or "It's none of your darn business!" If it's a very serious geologist, I say, "I'm working on a carbonate sediment problem in Majuro Atoll"—and that's true, though at so slow a speed that progress is negligible, a fact I don't mention. It's amazing how many people want one to serve on church committees and be an usher, give talks to retired and senior groups, serve on field trip and geologic meeting committees, run for political office, or help raise money for the poor starving Armenians. The answer to these non-dilemma requests is to travel extensively so no one can find you!

One soon learns that there is one major difference between retirement and work. When you are retired you can procrastinate with great efficiency, in complete contentment, without being conscience stricken. Put another way, the ideal retirement is one when you don't need to push for anything nor anybody nor to get any place, unless you wish to do so. The first requirement is to ditch the alarm clock! Branch Rickey, when told by some retired baseball players that they were unlucky, said, "Luck is the residue of design and desire." Yet a prevailing view about a bad retirement among many is that the end of teaching means freedom in poverty. That may be true for some but need not be so for anybody at all, provided plans for retirement are made early in life.

The crowning statement that really involves the important things about retirement was made by Thoreau: "It would be glorious to see mankind at leisure for once; it is nothing but work, work, work." The central word is leisure! In retirement "work" becomes another one of those dirty four-letter, Anglo-Saxon words of which there are so many.

Rocinante—
frequently our home away from home.

The Retired—A New Elite

There is a special group of retired people in the United States. Couples, with children grown, they travel in recreational vehicles as varied and individualistic as one can imagine. When they move south from northern climes, they're called snow birds. Driven home by hot weather in the south, they move as streaming hordes toward the north to spend summers in dedicated idleness. Many are full-timers, having abandoned their former homes and old friends to take up a lifestyle as knights of the open road. Disbanding sales follow a ritual now well-known to their children who have established homes of their own. Before the old home place is sold, their grown children gather around to acquire books, furniture, clothing, bed clothes, knick knacks, bric-a-brac, anything which the newly retired can't carry in their moveable homes on wheels, or do not wish to store against the time when many tire of a nomadic existence. Many inevitably will come back to some permanent base of operations later!

Most range in age from 55 to 65 when they make the momentous decision to travel. Some take to sailboats or power yachts, others buy pick-up trucks or double-cab trucks, drag along large heavy fifth-wheel trailers. Others go by motorhomes that range from 22 feet to 50 feet in length; many favor trailers pulled by large vans, trucks or heavy automobiles. All rail at the price of gasoline or diesel fuel, not realizing that fuel costs only four to six times what it cost in 1940, whereas bread costs 10 to 20 times as much as in the halcyon days of low inflation but equally low incomes. The new elite in motorhomes soon realize the lack of options for independent travel once they are hooked up at recreation centers for stays of weeks to months. So they buy small cars, light trucks, and even heavy cars and four-wheel drive cars to drag along on two wheel dollies, or with front or rear ends off the ground by special mechanisms designed to trail easily behind motorhomes. Others drag these utility vehicles along "four-on-the-ground," as RV terminology goes.

There are younger "retireds" who favor vans, gaily colored sin bins. Some carry tents and camping paraphernalia in small trailers or use tent trailers. In lessening numbers, one still sees retired people using truck campers, one of the most nimble of all recreational vehicles. The combinations one sees on the highways of America from Alaska to the Panama Canal are astounding! The retired elite, and not-so-elite, seemingly are forever on the move.

What made this travel revolution possible? When President Eisenhower decided, or approved the decision, to build the interstate highway system in the United States, the stage was set for a full-blown revolution in the travel of retired people—the new burgeoning of gypsies in America—in larger rigs than existed before 1954. The RV industry delights in publishing photographs of home-built rigs designed by tinkerers before 1954. (Of course the travel trailer has been in existence since the Conestoga Wagons were first created by the Pennsylvanian Germans in the early half of the 19th century.) Among the first of the lighter trailers, called house trailers then, was a popular model called "Covered Wagon." Trailed behind cars, these, after the early 1950s, led to mobile homes, ranging from mobile trailers to park models moved to relatively permanent mobile home parks and set on foundations. Inside almost all of these mobile home parks one sees stored smaller, more mobile units such as campers, trailers, and tent trailers owned by denizens who simply must roam once in a while from their solidly anchored mobile homes.

Some people go it one better by having two mobile homes, one in Texas, Arizona, California, or Florida, the other up north, with a third RV to travel in back and forth as the seasons change. It is interesting to note that retirees generally move along lines of longitude. For example, eastern Canadians, New Englanders, and New Yorkers go to over-crowded Florida; Westerners move from British Columbia, Alberta, Oregon, and Washington to Arizona, Southern California and Baja California. The ones from middle north America are likely to be all over the place, but many favor Mexico and the Rio Grande Valley of south Texas.

In long years past—notably the 17th to 19th centuries—no such mobility was possible, for roads were bad or non-existent. Many movements for generations took place via rivers, notably the Mississippi, Ohio and Missouri. But non-river people moved in successive generations via Indian trails and wagon roads during the late 18th century, all of the 19th century and the early part of the 20th century, the migrations remarkably parallel to lines of latitude. This amazing movement from generation to successive generation saw Eastern Canadian Tories move to Saskatchewan, Alberta, British Columbia, and even on to Alaska. The Pilgrim and Puritan ancestors of my wife moved from Massachusetts and Connecticut westward to New York, southern Ontario, Ohio, and Michigan.

My own Pennsylvania Dutch ancestors moved in three generations from Pennsylvania to Ohio, Indiana, Illinois, and Iowa. Some of the more obtuse Mennonites moved to Canada, back to Ohio, down to Mexico, even to South America to establish farms and keep to their strong religious beliefs as bulwarks against excessive modernism. In both avoiding and in mingling with the "fancy people," these Pennsylvania Swiss now total well over several million in number and freely intermarry based on propinquity. At least a million of them still speak Pennsylvania Dutch.

The people of the Carolinas moved west to Tennessee and Kentucky, thence moved to Missouri, Arkansas, Oklahoma, and Texas in the mid-19th century. Many of the Huguenots moved to Mississippi and Alabama and Louisiana. Out of all of these original Easterners, north to south, a determined several thousand took to the roads and trails through the forests in great covered wagon migrations, which either funnelled through St. Louis to go to California at the call of gold or the Oregon Trail to the northwest. Some veered southwest to travel the Santa Fe Trail. All displaced Indians, themselves newcomers to America across the Bering land bridge during maximum glaciation 27,000 years ago.

America's people are a dissatisfied lot, always on the move. Is it any wonder that we are the most mobile people in the world who travel for almost any reason at all? Our ancient background has been that of leaving safe havens in Europe to cross the sea to find new homes for many, many reasons. It's in our blood, and the retired elite express the freedom so urgently sought by people who live longer, have been more successful financially, and continually seek to avoid severe climates after years of work, saving and investment. They are a hardy lot! Younger people rail at them because of social security taxes and may see the retired elite as parasitic on society. I personally revel in the freedom I have earned to join this select group of millions through 56 years of school and work in several careers. Long may we travel!

The Put-Down

Very few scientists, active in endeavors of all kinds, escape put-downs by various types of individuals. Many such encounters are vicious, based on jealousy, scientific greed, and efforts by the perpetrators to elevate themselves above people more capable than themselves. The put-downs below were all done in good humor by friends, and they all happened to me.

Professor Marland Billings, Harvard Professor of Structural Geology in the Division of Geological Science for years, had many of his doctoral students work on the complex geology of New Hampshire. I took several courses with Dr. Billings, and tough taskmaster as he was, he had a cutting type of humor, and even could laugh at himself from time to time. He published complex diagrams in his articles for the bulletin of The Geological Society of America. One of his most penetrating articles on mapping ring dikes, crosscutting sills and dikes invaded by granite intrusions—all involving metamorphic terranes—led him to devise all these relations in one block diagram. Before sending this particular article off to GSA, he handed the block diagram to me, as one of his students, and asked, "Would you please explain what you see in this diagram?" I examined it with care and told him the sequence of geologic events as he had it depicted. Obviously pleased, he then said, with a mischievous twinkle in his eye, "I figured if you could understand this diagram, anyone could!"

I was at a scientific meeting involving geologists in El Paso, Texas. One of our former doctoral students, Dr. Paul

Lambert—on whose dissertation committee I had served at the University of New Mexico—was standing with a group of young geologists. Paul was, at the time, a professor at a Texas university, and I had been retired for several years. I didn't know the lads talking to Paul, and I wanted to say hello to him. As I walked up, Paul said, "Gentlemen, I want you to meet one of my professors. He is Dr. Wengerd, a professional ne'er-do-well!"

The best advice that involves research and reward toward retirement is to leave when it's time to quit. Better yet, let go as soon as you can. The simple retirement ceremony, where university president Bud Davis gave each of us a handshake and an award scroll, was followed with a visit among the 42 people who retired in June of 1976. Afterwards, academic vice president Chester Travelstead, a friend, said to me, "Sherm, I don't know why we made such a fuss about your retirement. You've been retired ever since you came to this university!" My retort: "It wasn't a fuss and not much of a ceremony!" Dr. Travelstead had much earlier heard me claim that I began to plan my retirement when I was a Naval officer during World War II.

Many other interesting put-downs have occurred to me virtually all my life; I seem to attract put-downs for reasons I can't fathom. One of the most interesting occurred at a symposium assembly in my honor by the department of geology at the university in 1990, 14 years after I retired. The following chapter, titled "Dr. Black Lays It On," is a condensed version of the put-down section of what Dr. Bruce Black said that evening. A few months later Capt. Black was appointed Rear Admiral in the U.S. Naval Reserve.

Dr. Black Lays It On

"When I was asked to host this evening's activities, and especially to help with the honoring of Sherm, I was quite hesitant to do so as there are others who could do a much better job, who knew Sherm better, and whose lives and experiences more paralleled his. But then the more I thought about it the more I realized that this was an unparalleled and God-given opportunity. I come here tonight to praise Caesar, not to bury him. But you must understand that in 1962 when I came back to graduate school I made all A's—that is, except for one class! Now you're beginning to get the picture. Sherm gave me the only B of all my master's and doctorate work and, therefore, I thought, ruined an otherwise perfect academic record. Now, Sherm, it's my turn! You have no alternative now but to sit there and squirm in your chair. To start my research into revenge, I began by comparing your history and the things you had done with my own. To my surprise I found that after all these years we, indeed, have a great deal in common:

- Both of us studied to be geologists, although I got my good grades the old-fashioned way and earned them, while you married a professor's daughter.

- Both of us went to work for Shell Oil Company and earned our reputations as oil finders with the petroleum industry.

- Both of us enlisted in the Navy, and both of us eventually rose to the rank of captain while working with aerial photo interpretation and intelligence related activities.

• Both of us taught at UNM...I for only a few semesters as Distinguished Professor of Petroleum Geology, and you for years and years and years, I presume because they made you keep doing it until you got it right!

• Both of us love aviation, and we are both fliers. You having hundreds of hours in single engine aircraft. I have hundreds of hours in single engine aircraft also. You must admit, however, that I have outdone you in the number of forced landings and slight incidents (read that as significant crashes) which we have both been involved in and walked away from over the years. In every case mine involved unavoidable mechanical failure in which I guided my stricken aircraft away from imperiled civilians, sacrificing my own self- interests and body for the protection of the innocent. While you, on the other hand, undoubtably had your problems due to pilot error and were very lucky to just get out alive.

• Both of us have built our own aircraft. You were the recipient of national awards for your model aircraft building skills. While I have received no national awards, I did get great local and statewide TV and news media recognition when I dumped my homebuilt Spitfire on its nose on its maiden flight—even though it was a perfect three point landing: two wheels and the nose. Several incredulous witnesses to this aerial feat led people to ask me who the hell taught me how to fly? I, of course, immediately told them it was you. They understood.

• Both of us married beautiful women and have lived with them in perfect bliss ever since. Although I'm not really sure how you ever managed to trap Florence; it is a wonder how she ever put up with you.

• Both of us have been active in geological organizations and associations, with you rising to the pinnacle of success as the president of the American Association of Petroleum Geologists, while I was only the president of the Rocky Mountain section of that great organization.

Now while I made this comparison, it became evident to me that everything you did I seemed to have done also, in some form or another...and usually faster and better. At last I have caught up with you. And tonight here we are, and this is THE hour I got to put you under the gun for all your past sins."

(University of New Mexico, Honors Symposium Banquet, September 29, 1990. This "Roast" is published with the permission of Dr. Bruce Black, president of Black Oil Company, Farmington, New Mexico.)

Honors Symposium Banquet
September 29, 1990

Mom Wengerd
(fourth from the left)
with her brothers and sisters.
c. 1935

Part Four

UNFORGETTABLE PEOPLE

Temptation
Enthusiasm
The Wanderers
Freedom, Liberty and Mediocrity
The Self-Actualizers
So Near to Fame
Lewis and Al
Martha Belle and The Pig Stand
The Scout Check
Tommy Knockers

Temptation

Oscar Wilde once said, "I can resist anything but temptation." If trouble is indeed a simple definition of being human, then so is temptation. Why fight the delicious feelings of temptation? Our Christian ethic is replete with warnings that one must fight temptation at all cost, or failure will be our lot. Let's take another tack. What is there to look forward to were life bereft of temptations that give rise to ambitions?

My father and I were walking along in Times Square Saturday evening many years ago. It was long before it became a honky-tonk part of New York City with drug dealers, pimps, hookers, gangs with knives and guns, and general riffraff. We were thirsty, so we yielded to the temptation to go into a bar for a beer. If we had fought off the temptation, we would not have met a pudgy, roly-poly man smoking a long expensive cigar. He hailed us as he was rearranging some decor in his newly opened bar, and we recognized him as Tony Canzonari, despite his long retirement from boxing. Tony had been the lithe, fast and punishing light-weight boxing champion of the world. Many years before, we had seen him fight in Canton on his way up to the crown. We told Tony this, and he beamed, but not enough for him to stake us to a beer. Not to mind. We enjoyed his salty conversation. My father, an avid student of boxing, had seen all the great fighters and knew Jack Dempsey personally, and his reminiscences were a delight to hear. Had we not relented to our temptation, we would have missed an interesting experience.

Bill Breed, that peripatetic friend of mine who acts as guide to groups wishing to explore out-of-the-way places, believes that virtue has its own punishments. He also recognizes that one man's perversion may be another man's diversion. To his credit, when a temptation strikes him, he embraces it and gets himself into some of the most amazing pickles of any man I know, maintaining that if he's not in hot water most of the time, he's not happy. G.H. Chesterton said, "I like getting into hot water. It keeps me clean." Obviously he was a man unafraid to take a chance. Serendipity is not a planned condition; to savor life one must be tempted to allow things to fall where they may. It adds to the thrill to accept temptations as opportunities for adventure. Frances Rath believes that experience is not so much about what happens to you as it is what you do with what happens to you. The Reverend Rath, our Congregational minister, is subjected to a powerful temptation—just as I am—three times a day. We love to eat! Some years ago he went on a crash diet, fought temptations to eat, lost lots of weight, became lithe, and his sermons became dull, lifeless admonitions straight out of the Bible with little fire. When he recovered his senses and charged into meals like there was no tomorrow, the whole church lit up, and his sermons were no longer hell-fire and damnation but marvelous essays on how great life really is, and what we can do to make our lives better.

There is nothing so dull as an ascetic afraid of temptation. Women who utterly resist all temptation become slatternly hags—dull, colorless bags of bones with never a joyful thought or action. They may always be right but are also miserably alone. Temptations are good for the soul. Musing about them leads to great dreams, and it is well known that dreams often turn into action with wholly unexpected results. Oscar Wilde, a man who yielded to gross temptations and wrote about them, once warned, "When the gods want to punish you, they answer your prayers." My interpretation of that fatuous phrase is that if you seek to satisfy a yearning, submit to a temptation willfully, fate (the gods) will make you pay for it. Ridiculous! Baryshnikov, that superlative dancer and come-lately comedian, put it this way: "Learn for a century and you'll still be a fool." How boring to continue learning and never do anything. Sir Peter Scott, a very serious scientist who had a sparkling sense of humor, wrote a book titled *Quotes, One Liners, Funnies, and Groans*, a compilation of 639 sayings gathered throughout his too short life of taking chances. Once, after a lecture on temptation, he was congratulated by a lady who said, "We need more speeches like yours. They get us through the winter." His reply was "Everybody is somebody's bore, but I'm not sure whether it's better to be late than never." This non-sequiter was his yield to a temptation involving an effusive woman.

Temptation, not easily turned aside, should be welcomed if the results do not hurt anyone. To question the wisdom of an action based on momentary temptation should not be in the purview of untempted people. Lives there a man so wise as to be above temptation? To criticize a person for yielding to a pleasure marks the critic as unfeeling; perhaps ex-Senator Rudman, an ex-Marine, had such in mind during recent Senate hearings involving the Iron Curtain activities. "Infallible wisdom is not yet a trail found among mankind." If it weren't for lively temptations, ambition would wither and no progress would be made. Martin Mayer said, "Fear disables excitement." So, we raise our glass to toast temptation. May it long last and not get us into more trouble than we can handle!

Mr. and Mrs. Branch Rickey
Summer 1932

Enthusiasm

He was a man of boundless enthusiasm. Born in 1452 in Italy, he died on the second of May in 1519 at Chateau de Clous. Filled with honors, his facile mind invented machines unheard of before and not built until centuries later. Artist, artisan, craftsman, poet, and painter, he was revered by his student Francesco Melchi who, by the death bed of his mentor, said, "It is not in the power of nature to produce another such man." This man, whose mind ranged far beyond that of any previous man, was Leonardo Da Vinci. He himself philosophized, "I thought I was learning to live, but I was only learning how to die."

What is it within such a combination of genes that yields unheard of inventions? The vital spark of enthusiasm! From the ancient Greeks we get *en theos*, which really means "filled with God." So there must be something Godly within humans that creates enthusiasm. Branch Rickey said, "I prefer the errors of enthusiasm to the complacence of wisdom." Although enthusiasm and wisdom are not mutually exclusive, who has not thought at times "His enthusiasm is not guided wisely"? Enthusiasm is catching; it inspires deeds and thoughts foreign to dolts. To have it is to have visions of greatness. Few complacent people ever enjoy the bright shining gold of enthusiastic discovery. Though discovery may be tempered by patience, Saint Augustine prayed, "Lord, give me patience, but not yet!" Without enthusiasm our hopes could sink into the ooze of apathy, where they would die, as the Reverend Francis Rath believes.

It is true that one learns by observation, but association with an enthusiastic person is taxing for ordinary people. Serenity is not engendered by close contact with enthusiastic people, yet serenity is not apathetic to enthusiasm. Billy Wilder observed, "I'm too aware of the thin line between serenity and senility." I've never yet met a senile person who is enthusiastic. Alfred Lord Tennyson maintained, "I'm a part of all that I have met." If one associates with people old of mind, lacking in imagination and lacking the desire to learn new things, that's what one becomes: unenthusiastic. Might as well die and be done with all the hassle!

Enthusiastic people admire each other and many times "feed" on each other's dynamic outlooks on life. Unenthusiastic people consider them naïve, unsophisticated, unrealistic, and boring Pollyannas—best to avoid such dullards. One can judge a person by his attitude in the face of happy responses. The enthusiast typifies the iconoclast Ambrose Bierce's comment that "Admiration is one's polite recognition of other people's similarities to oneself." Samuel Butler must not have known many enthusiastic friends when he said, "All of the animals except man know that the principle business of life is to enjoy it." I've not yet found an unhappy enthusiast, but every dullard I've met was unhappy, was railing against bad luck, fate, lack of easy gains, and was the possessor of unflinching ennui. Creators of their own unhappiness, the unenthusiastic remind me of Abe Lincoln's comment that "Most people are about as happy as they want to be." Hopeful to an unnerving degree, the enthusiast is an optimist who finds his beer glass half full, not half empty. Enthusiastic optimists maintain that we live in the best of all possible worlds, whereas the spoil-sport pessimists believe that optimists may be right!

Stefan Zweig, in his 1946 book *Quotes from Balzac*, observed, "When [Balzac] lived, it was not to deceive but to indulge his exorbitant imagination and his sense of humor." Honoré de Balzac was the epitome of an eternal enthusiast, a true self-actualizer, who, like all such, enjoyed women. Zweig went on to describe Balzac thus: "He possessed the childlike good nature that we attribute to giants, and nothing could shake it." Truly a feather in the cap for naïveté and wonder, an optimistic enthusiast directed toward life in general. I have never met an unenthusiastic person who had a sense of humor or ambition. In all social intercourse, humor is the leavening that makes liking people possible. The statement by Will Rogers—"I never met a man I didn't like"—shows the power of a strong sense of humor, for liking everyone one meets is difficult indeed! As for ambition of the enthusiastic individual, it's good to remember Mark Twain's advice: "Stay away from people who try to belittle your ambitions. Small people will do that, but the truly great will make you feel that you too can become great."

The Wanderers

They wandered the roads of Ohio between the years 1919 and 1939—those were the 20 some years when I noticed them. Our home was on a dirt road between Berlin and Winesburg, not gravelled until 1920, but totally paved by 1922. Today it is U.S. Route 62, a major highway only parts of which run along the Interstates with other numbers. The southwest end of Route 62 is in El Paso, Texas.

The wanderers were always men, and it seemed they always came from the north. Men out of work, many widowers, and many without permanent homes, but all of them hungry and carrying bundles of what, no one knew but themselves. Many stopped at our home, timidly knocking on our back door, asking if they could do a bit of work for a meal. If it was near the end of the day, they'd ask if they could sleep in the hay mow of our barn for the night, always leaving cigarettes or a pipe plus matches with my mother to prove they would not smoke while bunked down in our barn. Mother almost always fed them; sometimes she had a little job for them to do. Much later I was to learn that our mail box had a little "X" marked on the wooden post, the almost universal sign that "here lives a kind lady" who will feed you with or without working.

My father called those men "knights of the open road." Few were derelicts, none of them ever stole anything that I know of, all were cheerful, and many were good conversationalists. They always travelled on foot in the summertime. When I was a little boy, their poverty struck me as an unkindness on God's part to allow circumstances such as this to exist. I was home every summer until the age of 11, so I got to know some of these road people, whom many in our community called tramps. Some unkind neighbors would call them bums. Most of us reserved that term for the men who "rode the rails." Once in awhile one of those bums would appear. One could tell they were not the usual tramp, for they were always dirty. Yet my mother would feed them but never let them stay in our barn. She knew that they were apt to steal some of our chickens. They would sack them up with heads pulled off and take them to a well-known hobo jungle by the Millersburg railroad that ran north-south through the Killbuck Valley from Wooster toward Columbus.

At times my mother, a devout Mennonite Christian woman, would sit and talk to them, with me sitting by, all ears, to hear the stories these migrants told of their lives. My mother would always query them about where they lived, whether they had families, what kinds of work they had done—all of them had at one time or another. The stories those men told were heart-rending. Some of what they told us may have been fabrications, but my mother appeared to take them at their word—a real lesson in human relations. If they didn't ask for money and if they offered to do some work, after they ate soup and a sandwich, my mother would give them a small amount of money and send them on their way. Naïve perhaps, but that was my mother's Good Samaritan act for the day for a total stranger.

This was the time of the 1920-21 recession, before the time of unabated prosperity and what seemed like a false plenty from 1923 to 1929. After 1929 people lost their jobs. The smoke no longer came belching out of the stacks of Republic Steel in Canton and Massilon. People lost their homes, their savings, their families in the cities of Cleveland, Akron, Youngstown, Canton, and Columbus. Soup kitchens and food lines from 1929 through 1940 couldn't feed everyone. Many men hit the roads into the countryside where Pennsylvania Dutch farmers hardly knew that a depression was taking place. It was these wandering men who impressed me greatly about the rigors of poverty. Yet once in a while one would come along—usually a scissors sharpener with a small tool box. He would sharpen all my mother's scissors and knives. But she would notice that these scissors-sharpening tramps carried a small bottle of cheap wine in their tool boxes. To those, though she would feed them, she would give no money, or if they asked for money because they file-sharpened scissors and knives, she would pay but not feed them. Her methods taught me much about psychology and business.

One day in June of the summer of 1925, a neat older man came to our door. My mother was not home. He had on a frayed uniform coat with some tarnished gold buttons missing. He told me he had lost his job as a security guard at the Cleveland library. I made him two sandwiches, and we visited. He was obviously a cultured gentleman, far down on his luck. I had been saving rigorously to buy firecrackers for the Fourth of July. He asked if he could do some work, didn't ask for money, and thanked me for the food and the visit. As he started to leave, remembering my mother's charity, I said to him, "I've saved up $1.62. You can have it." He took it, thanking me profusely, and when firecracker day came along, I didn't miss the fun because my father bought the firecrackers for us boys that year.

These wanderers of the open road were walking lessons in poverty so impressive that I vowed with all my might to avoid poverty and try to arrange my life so I could help others who really need it. Then I remembered something my successful father told me: "Whatever you give from the heart, you will get back ten times over." To me that seemed to be a great deal. And it works, though that's not what charity is about!

Florence and me at
Cape Spear, Newfoundland
August 1972

Freedom, Liberty and Mediocrity

On April 10th, 1978, at a national meeting of the American Association of Petroleum Geologists, Ronald Reagan, then governor of California, gave a luncheon talk in San Francisco. His talk merits notes on his philosophy. These are exact quotes from the speech he gave then without notes or prompter. I record them here without specific comment about each one.

> "Profits, property rights, and freedom are inseparable; you can't have the last without the other two!"
>
> "Freedom is never more than one generation from extinction."
>
> "Business doesn't pay taxes; people pay taxes."
>
> "Our problem isn't a shortage of funds; it's a surplus of government."
>
> "Where were you when freedom was lost?"
>
> "If you go to bed with the government, you might get more than a good night's sleep."
>
> "Geologists are an endangered species."

Throughout Gov. Reagan's talk, his enthusiasm flooded the room. He literally bounded onto the platform when he entered the room filled with thousands of geologists and their wives. His manner was upbeat, his diction perfect, the cadence of his pronouncements dramatic. In the later years when he was president, he was the bane of the liberals. He always wished to change things whether the status quo was correct for the times or not. After he was president, he again addressed our association with vigor, despite a terrorist threat on his life. Reagan correctly assessed the vast increase in the ~~in June of the summer of 1925.~~ national debt as being the fault of Congress and so was his lack of the line-item veto to cut their spending aimed at buying votes. In the last year of his presidency, the United States became a debtor nation. Although no liberty was lost, some freedom was lost, and no recession had yet gripped our spend-thrift nation. The printing presses rolled on and on, and the euphoria of growth led to a kind of spell-binding mediocrity encouraged by the liberal Democrats. Among those were ultra-liberal preservationists who would throttle free enterprise in these times. As Michel Halbouty put it so correctly and bluntly, "We're supersaturated with those fanatic sons-of-bitches."

A. L. Rouse, an English historian, said in 1961, "The longer one lives, the more one sees that thin ~~aadzevbxcvbnmgs~~ are really as silly as they seem." But there may be hope, as noted by Keith Jackson, a TV commentator: "It's funny howwisdom seeps in as the years pass by." The search for freedom and a deep sense of the necessity for self-determination are what drive a man to be a pioneer in any risky, daring, adventurous endeavor. A free government must encourage the search for liberty despite a crass and unfeeling bureaucracy. Why are we now so submerged in red tape that limits our freedom, cudgels the very essence of liberty, and leads us all to be reduced to faceless mediocrity? Is it the lawyers, the clerical mind, the socialistic efforts to bring every one to the same low level of achievement? Pogo was right: "We have met the enemy, and they are us."

Heraclitus, that venerable Roman philosopher, said, "Nothing endures except change." So why can't change always be for the better, given the intelligence of the American people? Abraham Lincoln, a Republican president sometimes claimed by the Democrats, said, "I believe each individual is naturally entitled to do as he pleases with himself and the fruits of his labor, so far as he in nowise interferes with any other man's rights." Have we as a people lost this wisdom and strait-jacketed the free mind that makes for progress? Doris Fleeson must have considered the lethargy of group actions that stifle initiative when she said, "Against stupidity, the gods themselves struggle in vain." Many lone spirits in the wilderness cried out against the inhumanity of man to man when Jeremiah, who suffered two exiles, asked, "Why do the wicked prosper?"

If we are to believe as Christopher Morley did—"There is only one success: to be able to spend your life in your own way"—then we must seek to banish all strictures from government that limit the flight of the human imagination. I am reminded every day of two of the greatest bits of philosophy one can know. Amos Fortune, a slave in Virginia, educated by his master before he was freed, said, "Freedom is uselessly wasted on the ignorant and the indolent." Ben Franklin, noting the dangers foisted on the people of France, both by the Royalists before 1792 and the Revolutionaries after 1792, believed, "Where liberty dwells, there is my country."

Without freedom of choice, and the liberty under law to pursue a course of action, little can occur but mediocrity. If mediocrity owes its origin to stupidity rather than to ignorance, not much progress is possible. Where mediocrity is dictated from above, intelligent people will rebel, and if enough do, the fat's really in the fire. Isaac Disraeli, father of Benjamin, said, "It is a wretched taste to be gratified with mediocrity, while the excellent lies before us." Napoleon maintained that "There are limits to rascality, but there are no limits to stupidity." Obviously when both are dominating a Congress, a country is liable to go to hell fast. Albert Einstein was fond of saying "Great spirits have always suffered attack from great mediocrity." Nowhere is this more evident than in hearings run by the Senate Judiciary Committee charged with examining the qualifications of a president's nominees to the Supreme Court. It is astounding that so many on that committee—which almost automatically calls the kettle black, when most of them could not possibly withstand deep scrutiny of their ethics—will vote a straight party line of nominees that are known to be qualified. Of many of the critics of a nominee in recent hearings, Senator Orrin Hatch said of one nominee, "Your actions speak louder than their words." This statement mystified at least two rather stupid committee members who shouldn't have been in the Senate at all, so Senator Hatch explained it to them, to the chairman's disgust.

Any group in the United States in the forefront of guarding freedom and liberty is only itself free if it cherishes creativity, individuality and originality in all people for the good of everyone. Where overprotection of the incapable is rampant in any land, where failure is not allowed, where ambition is choked by burdensome laws, there liberty does not reign. Autocracy and dictatorship may flourish, or anarchy and pandemonium will rule to the detriment of progress.

The Self-Actualizers

It happens to women and men—a view of life as something to be lived to the fullest. They are born with an earthy love of body, especially the life processes. Most of them, undeterred by convention in their personal lives, adapt easily to the society of their times. They have strong elimination and circulatory systems, great love of food, very strong sexual drives. Surprisingly, all self-actualizers have within their psyche a bit of voyeurism and narcissism—some much more than others, but it's there. They are not shocked by any of the body's effluents. In fact, they revel in recognizing them as the results of good health. They have amazingly active saliva and lymphatic juices, unusual in average human beings, and they have the self-disciplined love of complete relaxation. They are joyful, have lots of fun, love to travel, are fond of alcohol but under careful self-control. They are the doers and movers in our society. No nation claims more of them than any other; no race is predominant in their percentage of the population; and most of these people have no racial hatred. Above all, they are capable of working very hard, undeterred by their surroundings. You may think I'm describing genius. Not necessarily, but many are just that. The above litany describes the self-actualizer. If you are one, you realize it very early in your life.

I am indebted to Professor David Benedetti, friend and retired professor of psychology who taught for years at the University of New Mexico, for introducing me to the works of Abraham Maslov. Look around you; there are more of these people than you think. Not all are eminently successful. Ben Franklin once said, "If you would acquire property, first acquire a good wife." Some of the self-actualizers I know did not marry well. Too interested in dissecting the world around them, some are persuaded by their strong sexual drive to marry too early and not well. Such capable people do not walk in fear. They trust everyone until crossed. Then they are doubly vindictive with those, yet overlook slights from lesser acquaintances. Most have strong personalities. All of them have fantastic vocabularic abilities. Many are blinding orators unafraid of criticism. Some are engaging writers who can turn a phrase about common things and situations into brilliant, cogent, unforgettable prose or poetry. The greater number by far are honest, not needing to be dishonest. Those who vastly exaggerate, bend the truth and emulate warlocks are usually found out, end up in prison or commit suicide. All know who they are, and they are quick to recognize fellow self-actualizers. Few have patience with stupidity of the common herd, and they avoid people from whom they can learn little. Do not get the idea that these people are commonly assessed as paragons of virtue. Most men and women and religionists are loath to give self-actualizers any credit at all.

Who are some of these self-actualizers? If it is true that it takes one to recognize one, that applies to thieves as well. Rummaging around in literature and in history, one finds many examples. Here are a few, some listed by several authors: Napoleon Bonaparte, Abraham Lincoln, Ralph Waldo Emerson, Elbert Hubbard, Henry Ford, Kirtley F. Mather, Thomas Edison, Robert Burns, George Washington, Walter Chrysler, Arthur Compton, Harvey Firestone, Woodrow Wilson, Thomas Jefferson, Booker T. Washington, Andrew Jackson, Napoleon Hill, George Bush, Alexander Hamilton, Winston Churchill, Robert Millican, Albert Einstein, and William James. If you look far enough back in history, one must include Jesus Christ, Alexander the Great, Mohammed, Charlemagne, Moses, Abraham, Plato, Aristotle, and Ulysses, plus a host of other people unrecorded in history, lost in the mists of time.

Kirtley F. Mather—
A self-actualizer

Can one follow the suggestion of the psychologist-philosopher William James and seek to become someone better by acting like someone better? Doing as well as possible, difficult things that we detest and finally liking the challenge? Can we become what we want to be? Despite the generally held notion that we are born with a spark of the divine—or the red-neck view that "You've got it or you ain't"—I believe one can become a self-actualizer by acting like one, by learning new patterns of behavior through diligent application and deep thought. I know people who have done it! Shirking off man-dictated strictures and charging off on their own, I know of several people of average intelligence who have become self-actualizers, much to the surprise of their doltish friends! The Latin phrase comes to mind: *Aut inveniam viam aut faciam* (I shall either find a way or make one). Sidney Smith, perhaps unwittingly, described the kind of person who can never become a self-actualizer by saying, "He spent much of his life lowering buckets into empty wells and frittering away the rest of his life drawing them up again." To those of high capability, Ted Turner's admonition applies: "Either lead, follow or get out of the way." And finally, there is hope for all to attain a measure of self-actualization. A philosopher of the 19th century said, "Sow an act, reap a habit; sow a habit, reap a character; sow a character, reap a destiny." Maslow, that perspicacious student of human nature who coined the word "self-actualizer," couldn't have said it better!

So Near to Fame

Most people on the sidelines of life's playing field, filled with not usually known men and women, realize that their contacts with the "greats" are limited. At times not only does one just see them, but one may meet some of them, mainly by accident. Below are brief vignettes of people who are or were known to thousands of people through their accomplishments in many fields of endeavor. Many have had biographies written about them, some have penned their own life stories, but all of them have been profiled or have themselves written books and articles about their lives and times in the national scene. Some have been reviled, but all were what I choose to call famous, and I either saw them or met them. By 1994, most have "shuffled off this mortal coil," but some are still alive. Many have become legend and some have already been forgotten—the almost inevitable end of fame for most humans. The thrust of this compilation is not to drop names but to honor some of those of fame who have had direct or indirect effects on my own life.

Winston Churchill came to Washington, D.C. a number of times early in World War II. One day I noticed a huge crowd of people streaming toward a cordoned-off area east of the Capitol. Parking the car on the way to duty at the Navy department, I walked to the area, and there, standing in an open limousine, dressed in his war-time cover-all flak suit, was Churchill, both arms up in the air giving his salute, the two-fingered "V," he made so famous. He was on the way to address the joint sessions of Congress on his views of the war's progress in Europe.

Kirtley Fletcher Mather was one of the best known professors at Harvard. Although a geologist-geographer, he injected himself into the national scene by his activities as a witness at the Scopes Trial—the so called "Monkey Trial"—in Tennessee in 1925. Later he created a national stir by refusing to sign the Massachusetts Loyalty Oath, stating that no piece of paper would guarantee anyone's loyalty. He later recanted and signed the loyalty oaths as required. As Dr. Mather taught at both Harvard and Radcliff, he said, "I presume that makes me twice as loyal as most other professors." Professor Mather was labelled a "pinko," or a communist sympathizer, by Senator McCarthy and lumped with a group of other famous scientists and writers in an article in *Time Magazine* in the 1950s. His stand, and that of others who nurtured the anti-McCarthy campaign, led, in part, to McCarthy's death in disgrace.

Thomas Mitchell, well-known actor and one of the heroes in the movie *Shangri La,* came to the University of New Mexico to lecture before a huge crowd in the student union building. He arrived early, came into the cafeteria, and looking around, found that no one was there to meet him. He knew no one, so I asked him to sit with us so he could get a sandwich and a cup of coffee. He joined us, all smiles, and for a half-hour or more, he regaled us with humorous stories about his acting career on stage and in movies. Why no one from the university had squired him about, I will never know, but whoever failed we thanked for giving several younger members of the faculty a rare visit with a great actor.

President Franklin Delano Roosevelt came to Harvard University during its Tricentennial in the fall of 1936. He was riding in an open car during a heavy rain storm. Seeing a parade of cars on Massachusetts Avenue bound for Memorial Hall where he was to speak, I stood on the sidelines and waved to him as he waved to the crowd. He was then a healthy, vibrant personality with whose political and economic philosophy I thoroughly disagreed, yet it was still a thrill to see the President, the first one I'd ever seen.

In 1944 I was serving with a Naval air squadron at Barbers Point Naval Air Station on the southwest corner of Oahu. Everyone was asked by the station commandant to turn out in full uniform to line the parade route during Roosevelt's inspection of the Naval air station. As an Assistant Secretary of the Navy in World War I, Roosevelt was highly partial to the Navy. Dutifully avoiding the Navy duties in readying our mapping squadron to move west toward the war areas, we all lined up to wave and salute the President that day. We were shocked to see an ashen gray man in bright sunshine look straight ahead with sad eyes and barely acknowledge the plaudits of officers and men at attention. Little did we know that he would die a year later.

The Comptons—Karl, Wilson, and Arthur—were sons of one of the premier families connected to the College of Wooster; Arthur, a famous physicist in atomic science; Wilson, a wealthy tycoon who became president of Washington State University; Karl, the well-known president of Massachusetts Institute of Technology. I met all three at various times while I was a student at the college between 1932 and 1936. All three were gentle, great men, who loved Wooster and visited when they could with faculty, administrators and students, and talked to us in Chapel.

The College of Wooster in the depression years of 1932 to 1936, as well as years before and after, invited many noteworthy characters to speak at Chapel, where attendance was required each school day. I remember particularly Carl Sandberg, born on January 6, 1878, speaking about Abraham Lincoln and playing the guitar, accompanying himself in song. The students gave him a standing ovation, and he spent some time after Chapel talking to some of us bold enough to accost him. He came up with several of his own well-known quotes to our delight: "I expect nothing, so when I get something, it's just so much velvet"; "Be good. If you can't be good, be careful"; and "Someday there will be a war, and no one will come." Sandberg later used these quotes in some of his writings.

Edwin Markham

At one memorable Chapel meeting, President Charles Wishart introduced a short, stout man dressed in black, wearing a wide-brimmed hat, like a cowboy hat. He took his hat off with a flourish, laid it on the podium and, without notes, entertained us with forty minutes of his exquisite poetry. The man was Edwin Markham. Afterwards, several of us had a thrilling conversation with him, and I took numerous photographs for which he graciously posed, hat on, smile on his face. Little did I know then that an agent of his who handled his business affairs had cheated him out of most of his estate through mismanagement, benign extortion and outright embezzlement. This was one of the reasons this old gentle poet was on a lecture circuit to colleges and universities during the Depression.

At another Chapel session, the college brought in Mr. Salo Finkelstein, a mathematical genius who could add a long column of figures in thousands as students called out numbers. Some of us unbelievers, generally those of us who were not mathematicians, wrote the figures down,and when Finkelstein gave the answer immediately, we added up the long column. It took us 15 minutes, and Finkelstein was right on the money. He demonstrated several other tricks of memorization, which amazed even Professor Yanney, the top mathematician at the college. This was long before computers, but we noted that Finkelstein's shoes were untied and his socks were not a pair. So much for brilliance!

He was a stolid man with white hair, and he could make a violin sing as he did in Chapel one day. We were enchanted as we listened to Fritz Kreisler play the violin and reminisce about his life as a musician.

The man who dazzled us with his stories, his wild exaggerations which he made seem so tame, and the enthusiasm with which he moved and spoke in Chapel was author and raconteur Richard Halliburton, an adventurer of such note that we were discussing him and his books for months afterwards. He made writing and lecturing seem utterly romantic, and adventuring a career to be followed at all costs. Many of us who heard him that day were greatly saddened when he was lost at sea a few years later sailing a Chinese junk across the Pacific Ocean.

Dr. James Conant, president of Harvard, a brilliant chemist who had a hand in creating the atom bomb, called a number of us doctoral students to his home for tea, tastefully served by the delightful Mrs. Conant on a Sunday afternoon. He inquired about our graduate studies in geology, about our professors, about the relevancy of our courses, and the type of examinations and their fairness. We were impressed by his knowledge of geology and his penetrating and up-to-date questions about the future of geology as an academic discipline and its practical applications. There is an aura about brilliant men, yet a humility about the best of them. Some years later when he awarded Ph.D. degrees to us *en mass*, he intoned, "You have now joined the society of learned men."

In 1950 on a geological field trip in northwestern New Mexico, a group of geologists from all over the U.S.A. stayed a night at the posh El Rancho Hotel in Gallup, New Mexico. Sitting in a booth, a group of us were discussing the geology we had seen that day when a tall handsome man moved from a table over to our booth and introduced himself. "I'm Joel McCrae. We're making a movie west of town." After we said our flustered hellos, we asked him to sit down with us. He was dressed in a cowboy hat, buckskins and cowboy boots, so everyone knew who he was: one of the great cowboy actors of his time. He asked us all what our jobs were, and then he said, "Do you know Manley Natland?" Of course we all said yes, and I mentioned that he was a friend of mine and a nationally- known geologist-paleontologist. Then Joel said, "Manley and I have been friends for years. We went to U.S.C. together, and I see him every once in awhile." We talked to actor McCrae for two hours, and later we found that the movie he was making was one of his last before he went into the sportswear business and ranching in the San Fernando Valley. A wholly unassuming man, a fine actor and a very interesting gentleman. He died only a few years ago at the ripe old age of 86, active to the last day!

My brother Wilmer and me
at Pearl Harbor
1944

My brother Wilmer was a sailor in World War II wiring rockets onto F4U Corsairs at Ford Island, Pearl Harbor, while I was a mapping officer with a Naval aviation squadron at Barbers Point on Oahu. When our days of leave coincided, we played golf at the Oahu Country Club, drank Primo Beer at Waikiki, drove the roads of Oahu in a Jeep from the transportation pool, or lazed on the clean but barbed wire guarded beaches. One bright day we were on the beach outside the Royal Hawaiian Hotel, a Navy R and R facility at that time. We plopped down beside a man in a bathing suit with a towel over his head, his rather large freckled belly bare to the strong Hawaiian sun. Wilmer and I talked about home and what we would do if the war ended by 1949—that was the time table we thought would be about right. Slowly the man with the towel sat up, his towel dropping to the sand between his legs. I glanced over, then turned to my brother and asked in a very low voice, "How would you like to meet a very famous movie actor?" He said, "Sure. When and where?" I told him to look carefully to the right past my nose. He did so and almost lost his teeth. I was sitting right next to Spencer Tracy! I introduced myself and Wilmer, and for a half hour had a momentous visit with him. He was in Hawaii visiting hospitals on a tour to cheer up Marines, Navy and Army men who had been wounded in the Saipan invasion. Before Wilmer was assigned to his duties on Ford Island, he was in the Navy pool of enlisted men and, in one of his most harrowing experiences, had helped off-load the Saipan wounded from hospital and other ships. Spencer Tracy wanted to know all about our various experiences in World War II to that point, and we told him how much we enjoyed his movies. Common as an old shoe, sorry he was too old to serve in the services, he showed to be as much a gentleman as he was a good actor.

Jim Arness and Jerry Colona, an unlikely pair of TV stars, made a tour of the U.S.A. in a comedy routine about their years after Arness began his career on TV as Matt Dillon in "Gunsmoke." I spied them at the old Albuquerque Sunport Terminal one Wednesday evening after I had been to a Naval Reserve meeting down the street on Yale Avenue, north of the airport. I went up to Arness, reached up high to shake hands—"He must be seven feet tall," thought I—said hello to Jerry Colona nearby, and began a tirade which left him almost speechless. A few weeks before, the newspapers carried an article stating that Arness was planning to quit "Gunsmoke." I enjoy any and all Westerns—movies, TV, stories, books, good, bad or indifferent—and found "Gunsmoke" to be very exciting, if not great theater (according to certain purists). I first told him how much I enjoyed his part as Matt Dillon, and then I said, "You can't quit 'Gunsmoke.' It's too darn interesting. How the heck can you even think of quitting?" Now on this particular evening I was in my Commander's uniform, and as he towered over me, I realized he looked, and was, much younger than he appeared on TV. He stammered a bit, thanked me, and we continued talking about his career. He laughed as he said good-bye. I could see the wheels whirring in the gentle giant as he walked away, wondering how he could be challenged by a guy a full foot plus shorter than himself. To this day I take a bit of credit for keeping Jim Arness on TV as Matt Dillon for years after that encounter.

Admiral Richard Byrd, whose son was a student of mine while I was an Austin Teaching Fellow at Harvard, was also a boyhood hero of mine because of his exploits as an explorer in cold regions and a daring user of airplanes. One day in 1945 before I was sent to Alaska, I was walking north from the old Navy Department, and a figure in resplendent whites, an Admiral, snapped off a salute at me, a lowly Senior Lieutenant, and I fell all over myself to salute him! As he passed smiling with a cheery "Good afternoon," I realized I had just passed Admiral Byrd.

One day in the late 1950s in Mesa, Arizona, I stopped at a Shell station for gasoline, and a tall handsome man greeted me with a hearty "Howdy." I recognized him immediately as Paul Dean, brother of ol' Diz Dean. Both were great baseball players, but Paul was nicknamed "Daffy" by news writers. Though not as famous as Dizzy, Paul loved to talk about his brother and baseball. As he serviced my car, we laughed about some of the phrases for which Diz became famous. In July 1974, ol' Diz died just three weeks after I saw him as guest host on TV's Monday night baseball game. All of us who played baseball, no matter how minor our contributions to the game, knew many stories about these two Arkansas hillbillies who added so much humor to the game. Most of us know that Diz went to some baseball players' heaven where all hits are home runs, or pitchers only throw strike-out pitches, and there are no angry coaches and nasty umpires, and no injuries. Diz once said, when asked why he bragged so much, "Well, it ain't braggin' if you can do it." His famous statement, "Hey that runner just slud into third base," has made many a baseball story. It was only through meeting his "little" brother that Diz lives on.

A tall dapper man whom I had never met was in the board room of a hotel as a group of us who were on the board of directors of Thompson International Ltd. met in 1960 to discuss strategy involving an unresolved lawsuit against our company, the ULA. Mining Company. General Robert Lee Scott, Sr., author of *God is My Co-pilot*, and 12 other books up to that time, was a much decorated World War II Army Air Force pilot in China. He had a bearing about him that foretold his success later as a jet pilot in the U.S. Air Force. There are some people one meets whom one likes immediately, famous or not, and in our discussions of company policy, he came across as a man anyone would enjoy knowing.

Books have been written by and about one of the great Arctic explorers of all time. He was a Captain, British Master of any size ship by the age of 26, born in Newfoundland, a naturalized citizen of the United States, a Lieutenant Commander in our Navy during World War I, but before that a companion to Admiral Robert Peary on his dash for the North Pole in 1909, Arctic Skipper and owner of the famed Arctic schooner, *Effie Morrissey*, teacher between 1926 and 1946 of many a lad on how to be a manly and tough sailor.

He was Captain Robert A. Bartlett with whom I shipped to Baffin Island on the *Morrissey* in 1943 during World War II.

Captain Bob Bartlett
on the *Morrissey*

Jack Kennedy was at Harvard as an undergraduate the same four year span when I was there as a graduate student. I saw him at various times in Harvard Yard as he was going from class to class, presumably. In the manner of those times between 1936 and 1940, Harvard men wore suitcoats, neat shirts and neckties—by whose dictum I never found out. But their trousers and shoes showed more of their true personalities, or so I thought. Disreputable pants of all kinds, all colors and conditions were *de riguer*, and shoes could be any kind. Jack always wore tennis shoes, and wherever he went he ran. I never saw him with the green cloth bag so prevalent for carrying books, notebooks, pencils, papers, slide rules, etc. He was tall, thin, intense, and obviously had an eye out for Radcliff girls. He showed no signs of the Kennedy wealth of his father. One Sunday, a group of students (friends of the Lyons family, who had five sons, one of whom was my friend Johnny) were invited to meet in a deep, hard-frozen Quincy quarry. There were many of these great granite quarries, mainly abandoned and full of water, the source of the famous Quincy granite widely used for facing stone, cobblestone, curb stone, and building blocks in New England. In that group of friends were Phil Reardon and Jack Kennedy. All were adept skaters, except me, but I had my skates along; everyone had skates in New England, it seemed. Sides were chosen up, someone produced hockey sticks, and a puck, and the game began. Reardon and Kennedy, good friends, were on opposite sides; Johnny and I were on the same side as Reardon. The game was fast, dirty, with heavy charging, blocking and sticks flying in all directions. I had never played ice hockey and I got run over, mauled, kicked around as befits a rank plebe among professionals. But it was fun, and the main battle was Kennedy versus Reardon. Reardon, a law student at Harvard, was later to have a high position in the government while Jack Kennedy was President.

Every month on a Thursday evening during the winter, the Boston Symphony would give a concert in the East Hall of Memorial Hall, that massive classic building that commemorated the Civil War and the Harvard men who died in that terrible conflict. The regular conductor was Serge Koussevitzky, but one Thursday evening he was replaced by Georges Enesco, composer of *Rumanian Rhapsodies*, and I was there in the balcony in one of the 50 cent seats Harvard students could buy. It was a glorious evening of thrilling music including one of my special favorites, "Rumanian Rhapsody #2."

Memorial Hall
Harvard University

One of the most unusual men I ever encountered was a Harvard professor of physiology, scion of one of New England's well-known families, a famous Arctic explorer, an Ensign in the U.S. Navy during World War I. Alexander Forbes, a medical officer, was lent to the Hydrographic Office of the Navy for Arctic exploration and charting. My contact with him first occurred in the Division of Air Navigation; in fact, he was directly responsible for my being in the Navy. In 1943 we were shifted to Chart Construction, and he was the senior leader of our Baffin Island-Frobisher Bay Expedition that summer. We were together on charting duties in a Navy unit in which I was Senior Line Officer as a lieutenant junior grade! He was a country gentleman, then over 65 years of age, and we withstood together the rigors of a season in the far North Canadian Arctic, charting sea lanes to supply our Arctic air bases during World War II.

Before our Navy Hydrographic Unit was sent to the Canadian Arctic, we were ordered to a conference at the U.S. Navy Department held by members of the Arctic-Desert Branch who taught us how to survive in that climate. Our instructor was the world-famous Sir Hubert Wilkins, a New Zealander well-known for his explorations in the Arctic. His instructions were in part responsible for the success of the Geodetic-Hydrographic Expedition that took place on Capt. Robert Bartlett's vessel *Morrissey* between August and October 1943.

Aboard the *Ambasador*
while circumnavigating Madagascar
Sir Peter Scott is on the right.
1986

Sir Peter Scott was the son of the explorer Robert Scott who died in the Antarctic on his unsuccessful effort to reach the South Pole ahead of Roald Amundsen. Sir Peter served as one of our scientific leaders on the ship *Ambasador* on the first exploratory circumnavigation of Madagascar under the aegis of Sven Lindblad. Sir Peter was founder, chairman and chief backer of the International Wildlife Fund, headquartered in Great Britain. He was a mild, unassuming expert on all wildlife. We found him to be an engaging instructor on animals and plants endemic to Madagascar.

Kirk Bryan
Harvard Professor

Dr. Kirk Bryan, Professor of Geomorphology at Harvard until his death in 1950, was a rotund, cheerful conversationalist. He made the study of land forms a science of almost unbelievable interest. Never at a loss for words and a teacher of many doctoral students who worked their dissertations in New Mexico and Arizona, Bryan was popular among faculty and students alike. Though he was not my doctoral chairman, he helped me directly with many new references and advice for my doctoral dissertation. A forerunner in geomorphology—which is the genetic offspring of the geographic science then known as physiography—his fame was worldwide. He had laughing eyes, was an incisive critic, earned his Ph.D. at Yale, and was a succinct author of outstanding bulletins for the U.S. Geological Survey, attracting many students who today are famous geologists.

There were other nationally-known professors of geology at Harvard at a time when it was the top graduate school in that science. Dr. Louis Caryll Graton was a millionaire economic geologist, dull teacher but brilliant scientist—truly a practical scholar. Dr. Donald T. McLaughlin, chairman of the Division when I was there from 1936 to 1940, was also a millionaire who became chairman of the board of Homestake Mining Company. Dr. Donald Leet was a geophysicist of renown who had been a student of Dr. Mather at Denison University. Dr. Percy Raymond, rotund lover of bourbon and good cigars, was Professor of Paleontology and an initiator of the new branch of geology known as sedimentology, my own academic field of study. Probably the most brilliant of my professors, Raymond was also the world authority on trilobites and an expert collector of pewter. Dr. Charles Palache was perhaps the best-known mineralogist of his day and co-author of several editions of Dana's *Principles of Mineralogy*; it was my privilege to study with him during his last year at Harvard. Dr. Alfred Romer, Professor of Vertebrate Paleontology, who was a popular lecturer and an engaging personality, was a man who had never taken a course in geology yet served one year as president of the prestigious Geological Society of America. Dr. Cornelius "Connie" Hurlburt, Professor of Mineralogy, was a student of Dr. Palache and co-author of several editions of Dana's *Principles of Mineralogy*. Dr. Esper Signius Larsen, great petrographer and initiator of the immersion method of mineral identification with the petrographic microscope, was a teacher who was so brilliantly disorganized that we learned his courses by trying to organize the helter-skelter of notes taken in his classes. Dr. Marland Billings was a structural geologist, fine teacher, tough scientist, blunt to a fault, and highly critical of fuzzy thinking. A master of intellectual power, he demanded excellence! Last but certainly not least, the most famous of them all was Dr. Reginald Aldworth Daly, Sturgis Hooper Professor of Geology, the highest paid member of the division who taught only one course: anything he chose to teach. Daly was one of those brilliant men who sought to unravel some of the most puzzling problems of the origin of the earth. His teaching was utterly methodical and a grand example of scientific method in the face of few facts.

Harvard Summer Field Trip
Canadian Rockies 1929
Pari Jas, Percy Raymond, Marshall Schalk
(left to right)

Tim Wengerd

There are two nationally-known experts of modern dance who were mutually dependent on each other for eight years before parting. The long and fantastic career of Martha Graham reached a nadir after she herself stopped performing. A fortuitous grant and renowned public support loomed on the horizon, and Martha Graham revived her dance company in New York City in 1972. She invited a young man to join her "new" company, a young dancer who was a founding member and a lead male dancer and choreographer with the Repertory Dance Theater at the University of Utah in Salt Lake City where he graduated summa cum laude in 1968. He was born in the Naval Hospital in Chelsea, Massachusetts, and, prior to his study at the University of Utah, began his dancing career at the age of 13 with the well-known Elizabeth Waters at the University of New Mexico. The Martha Graham Dance Company toured worldwide, and the young man soon became the lead male dancer of the company. Martha Graham created and choreographed difficult roles for him. With his help her fame was revived and was greatly heightened as he helped to make the company one of the most sought after in the modern dance field between 1972 and 1981. He was featured in articles in national newspapers and in national and foreign dance magazines. The company visited us in our home during a local performance, and we followed the company on parts of some of its tours. The young man was Tim Wengerd, my son, who went on to a successful career as a solo performer and choreographer until his untimely death in 1989.

Bradford Washburn, retired director of the Boston Museum of Science, became a special friend very early in my graduate student days at Harvard. He was instructor in the Geographical Institute at Harvard, established by Alexander Hamilton Rice. Rice, a stentorian type of philanthropist who was highly interested in geographical exploration, built the Institute buildings, and hired and paid the several instructors, one of whom was Washburn. The institute was attached to the Harvard Graduate School of Arts and Sciences. Several courses at this unique institute were qualified as advanced graduate courses in geology. He taught advanced exploration surveying and Arctic geology and very early in his life (1933) was made a member of the Explorers Club of New York. His work as a mountaineer—who very early climbed Mt. McKinley (Denali)—and his work on the glaciation of the St. Elias Range qualified him to teach courses at the graduate level, though he hadn't gotten his Ph.D. work finished as a student at Harvard. His exploits in air photography led to worldwide fame as a glaciologist in the study of modern glaciers, especially in Alaska.

Several groups of people are well-known in the musical field whom many people in the United States have heard and seen in their peripatetic careers as performers. Who can forget Jack Benny playing his violin with a symphony orchestra; Arthur Fiedler with the touring Boston Pops in Symphony Hall in Boston, in Albuquerque, and at Tanglewood; Andre Previn with the Pittsburgh and New York

Philharmonics; Serge Koussevitzky, conductor of the Boston Symphony; conductor of the Minneapolis Symphony Orchestra, Demitri Metropolis; Artur Rodzinski, early conductor of the Cleveland Orchestra; Hans Lange and Arturo Toscanini of the New York Philharmonic; incomparable composer Georges Enesco and violinist Fritz Kreisler, among many others who have filled many lives with beautiful music. I met Arther Fiedler three times, twice in Boston and once in Albuquerque. In Dallas at the Texas Centennial Celebration in 1937, I actually physically bumped into Paul Whiteman behind the scenes as I went back to say hello to Kathryn Grayson. Big band leaders at dances after concert performances—such as Tommy Dorsey, Sammy Kay, Kay Kiser, the singer Andy Williams with his touring group, and Benny Goodman with his orchestra—were among those I had the privilege of hearing in person.

Another highly visible group are members of our nation's Congress whom one meets if they are elected enough times and serve on important committees. I have met and talked at length with Clifford Hanson, past Senator of Wyoming, Hubert Humphrey, the late Senator of Minnesota, Senators Bob Dole, Birch Bayh, Robert Byrd, Pete Domenici, and Jeff Bingaman, and saw President George Bush while he was campaigning in 1980. Stuart Symington was with a group of politicos when I met him at the Capitol at an AAPG party. I've known every senator, representative and governor of New Mexico who has served since 1947—but that's no big deal in a state as sparsely populated as New Mexico.

Branch Rickey's home in St. Louis

Branch Rickey, one of professional baseball's most famous owners and general managers, was in his later career associated with the old Brooklyn Dodgers before they became the Los Angeles Dodgers. He was best known as the man who brought Jackie Robinson, a negro, into the Major Leagues. Robinson was the first black to break what had been an unfair and insurmountable color barrier. But Branch Rickey had been general manager of the St. Louis Cardinals earlier, and that is when I met him and his lovely wife at their country estate west of St. Louis. A boyhood friend of mine named Junior Rottman (son of the druggist in Millersburg and friend and fellow student of Branch Rickey, Jr.) and I went on a long trip into the Midwest to attend the Chicago World's Fair in 1932. Junior and Branch Jr. had been students at Ohio Wesleyan University, so Junior decided we should drive down to St. Louis in my car to visit the Rickeys. I had played high school baseball, so I was thrilled to be able to meet a man who turned out to be the most engaging personality I had ever met up to that time. We stayed at Branch Rickey's home several nights, ate meals there, ran around St. Louis with hard-driving Branch Jr., and were regaled with fantastic stories by the senior Mr. Rickey. In my sophomore year of college, I played on the Wooster team that trounced Wesleyan at their home field. Branch Jr. was the Wesleyan catcher, and, despite his repartee and heckling, I hit time and again the pitches he called for the pitcher.

I had no personal contact with Bob Hope, but probably no World War II veteran can forget his entertainment of troops in war areas. I saw him at EWA Marine base on Oahu during the summer of 1944, before our Naval Air squadron moved west to Kwajalein and Majuro atolls in the Pacific.

Early in World War II, during the fall of 1942, I had a long conference with Gene Tunney, the previous heavyweight boxing champion who had defeated Jack Dempsey in the "long-count" fight in 1926. Under Cmdr. Hamilton, Lt. Cmdr. Tunney was second in charge of physical fitness for the wartime Navy. His office was in the Naval Annex in Arlington, Virginia, as was the U.S. Naval Hydrographic Office at the time. I was a lieutenant junior grade, the first geologist to join the Navy as an Aircraft Intelligence Officer in the Division of Air Navigation. One of Tunney's fitness officers was Lt. David Hyde, assigned to duty at the Washington Receiving Center for enlisted men. Dave was the best friend I ever had during the four year span when I was at Harvard. The Naval medical board had sent him to St. Elizabeth's Hospital as mentally unfit to serve in the U.S. Navy—a shocking miscarriage of Naval justice. He had railed at the Navy, right up to the office of Adm. Ernest King, Chief of Naval Operations, at the U.S. Navy's insistence that every sailor going on leave be issued condoms. Dave was a born-again Christian and maintained that such issuance of prophylactics was an invitation to promiscuity. My conference with Lt. Cmdr. Tunney, a giant of a man, suave, imposing, and articulate, was unsuccessful as he pointed out to me that the Medical Board considered Lt. Dave Hyde insane. No amount of argument by me based on Dave's background and on our friendship while he was the Chief Physical Education Instructor at Harvard carried weight with

Gene Tunney. I had no sense of the Navy's rightness in issuing condoms to sailors. Perhaps it was necessary; no doubt it saved many men from cases of gonorrhea and syphilis. My argument was the unfairness and damage done to a friend's career. I kept up with Dave through the war years and later, and finally I heard that he went to Israel, wandered off into the Negev Desert, and was never heard from again. As I write this, we are in the Suez Canal passing along the west edge of the Negev, and I think of Dave Hyde's friendship and his tragic end.

Dave Hyde

One summer while I was a roofer for the A.S. Wengerd Slate Company, earning money to go to college, I was at my dad's warehouse to receive the day's orders. A tall, solidly built man, obviously a farmer, walked in to buy farm supplies for his large farm near Mansfield, Ohio. My father, knowing of my penchant for getting up early to write before going to work, said, "Sherman, here's a man you will want to meet." It was Louis Bromfield, a friend and client of my father's for many years. I had read some of his books. He wrote *The Green Bay Tree*, *The Farm*, *The Rains Came*, *Mrs. Parkington*, *Mrs. Smith*, *Pleasant Valley*, *Out of the Earth*, and *From My Experience*. Up to that time I had never met a nationally-known author. Bromfield was common as an old shoe, kindly, interested in my writing, and a genuine person—a real thrill to meet.

In 1972, because of my position as an officer in a national geologic society, I was on the Exploration Committee of the American Petroleum Institute attending an API meeting at a hotel on Nob Hill in San Francisco. There were many nationally-known oil company presidents and wealthy independent oil men and women there. I spied a little old man sitting off along one wall with no one talking to him. I went over to introduce myself and to chat. Surprised, he looked up, stammered a bit, then said, in his Texas drawl, "Ah, er, happy to make your acquaintance. I'm H.L. Hunt." H.L. Hunt, notorious for his wheeling and dealing, was one of the wealthiest men in America, based on his oil and gas holdings. We visited for about 15 minutes while he told me of some of the geologists who worked for him, his favorite food, strange exercises to ensure longevity, and some other arcane comments about people he didn't like. Only a few years later after he died did I realize that I had met a legend.

At that same meeting I again met Robert O. Anderson, an oil man whom I had met earlier as the founder of the Hondo Oil Company and owner of Malco refinery. He later became the chairman of Arco during that major company's maximum expansion as one of the largest oil companies in the world. Bob became a regent at the New Mexico Institute of Mining and Technology in Socorro, and today may well be the largest owner of private lands and ranches in the United States. He is a valued benefactor of the Anderson Graduate School of Management at the University of New Mexico. I knew him first as a junior oil man living in Roswell, New Mexico, where he lives today, not by any measure retired. He wears cowboy boots, dresses like any other rancher, is formal, though easy-going, and well-liked by all who know him.

The last brief sketch of famous people I have known is really about four men, no longer among us, two of whom had vivid impacts on my life. All are or were associated with aerial photography as well as many other technologies. Lt. Col. James Bagley was author of one of the first excellent books on aerial photography and photogrammetry. He was my instructor in that field at the Alexander Hamilton Rice Institute of Geographical Exploration at Harvard University. The courses I took with him and Bradford Washburn were the ones that led to my direct commissioning into the U.S. Navy early in World War II. A stentorian, formal, exact man, he had made his early reputation with the U.S. Geological Survey after service as an officer in the Signal Corps, U.S. Army, during World War I. He was a tough but fair taskmaster who allowed no mistakes, and I valued highly his friendship during my doctoral studies and later.

The second man of the four was Sherman Fairchild, inventor of the superb aerial cameras we used in World War II. He was the man who designed and built the well-known Fairchild 24 airplane, one of which I flew only once. He established an electronics company. A very rich man, he nonetheless paid rapt and formal attention in conversation and had ideas at the rate of "a mile a minute."

The third was Talbert Abrams, long-time owner of the Abrams Aerial Survey Company, one of the top air photographic companies in the U.S.A. He was also designer of cameras and progenitor of the special Abrams Air Photographic Airplane. His company flew air photographs over wide areas of the United States on contract to the U.S. Department of Agriculture Soil Conservation Services.

George De Garmo

The fourth man in this special technology, which was so much a part of my own air photographic career, was George De Garmo, owner and chief pilot of Standard Aerial Surveys in New Jersey. I first met him when his Naval air photographic unit flew the Frobisher Bay Project where I was Naval Charting Officer for the Hydrographic Office in 1943 during World War II. Later he selected me as the senior lieutenant line officer under him while he was founder and commanding officer of the Aerial Survey and Mapping Unit in the western Pacific. He was one of the finest commanding officers under whom I served during World War II. As our campaign of the Pacific Islands, mapping for the Hydrographic Office and ComAirPac Air Intelligence during World War II, was nearing an end, he kindly released me to become the Chief Aerial Magnetic Tracker and Photogeologist for Capt. William Greenman, Director of Naval Petroleum Reserves in our work on Naval Petroleum Reserve number four in Alaska, where World War II ended for me.

The thing to remember about such a long list of people whom I consider to be, or have been, famous, is that it's a subjective list. The names were chosen because I've travelled widely and lived a long time and never been too impressed by fame to keep me from telling these people what I thought of them or what they've done that affected me, and thanked them if they've bettered my life or blasted them if they've hindered our freedom. One is not easily forgotten if one has the gall to be afraid of no one.

Lewis and Al

They were true Westerners, this couple. Al had been married once before and had two sons. Upon the death of Al's first husband, she found Lewis Wright, a cowboy who ran a ranch on lease north of Woodward in western Oklahoma. In the proper time, Al and Lewis had a son named Samual Houston Wright, "Hoot" for short. But when Al and Lewis were first married, they lived on a large ranch covered with sage brush and prairie grass alongside the North Canadian River. My friendship with this stolid couple began in 1937 and lasted until the 1970s. Al, much older than Lewis, died first in March of 1978. During their lifetime they wandered from pillar to post, ranch to ranch. Later he worked for Lone Star Gas as a pipeline inspector. Then they moved to California where he contracted to collect garbage in a city outside of which he raised hogs. Contracts fulfilled, hogs sold, they moved back to a ranch outside Canadian in the Texas Panhandle.

After Al died Lewis stayed on that ranch through the good graces of a rancher who allowed him to live out the rest of his life retired. Lewis died in December of 1980. Lewis and Al never had much money, never owned a ranch themselves, always collected many friends, lived a full life, and spent all their money. Al drove like a demon; we called her "ol' lead foot." She was a heavy woman who loved to cook and run all over the West visiting friends and neighbors. Both were well liked by wealthy ranchers like the McQuiddys, on whose ranch they worked for a time with their friend, ranch manager Hub Hext.

Those of us privileged to know Al and Lewis pondered all these moves. Lewis was a splendid cowboy—short, squat, tough, and a lover of practical jokes who knew all the great rodeo cowboys such as the world champion, Bob

Al and Lewis

Crosby, who came to the Woodward rodeo each year to compete until his death in New Mexico in a Jeep accident after World War II. Al was earthy, bawdy and had a heart of gold. One summer when I was a weekend cowboy for Lewis, she cooked up a Sunday breakfast of pancakes and eggs after a rousing Saturday night of beer drinking at Martha Belle's Pig Stand in Woodward. I lit into those pancakes, knowing we would have a rough morning in the saddle moving cattle to a new river bottom pasture. The pancakes piled high with fried eggs, I bit into the stack, and the pancakes were filled with cockleburrs! Such raucous laughter you never heard before. Her comment to me, as Lewis rolled around on the bunkhouse floor laughing: "Well, you're such a prickly character, I thought I'd fix your kind of pancakes." Such practical jokes were her standard stock-in-trade.

When Hoot was born years later, we visited Al and Lewis, which we often did through the years as we travelled through Woodward or Canadian while I worked for Shell Oil Company. I asked Lewis one time, "Why at your age did you and Al have Hoot?" "Well," said he, "Al gets obstreperous every once in awhile, so this time I took her over my knee to paddle her, pulled down her pants, then I got a much better idea." Sex was a frequent subject. We used all the four letter words available in our Anglo-Saxon repertoire, and nothing was sacred but fun.

No holds were barred as I moved along in my career which they followed with glee. They came to visit us at Pueblo Buenaventura while I was teaching at the university. Their stock remark was "We knew you'd never amount to anything. Now look what you've gone and done, become a dern perfesser." One time in 1979 when we passed through Canadian, we stopped off to visit, and Lewis asked me, "Why are you goin' back to that dern East where you don't belong anymore?" I told him that my college was giving me the Distinguished Alumnus Award, the first in my college class. His instant retort: "You'd better git back there pronto before they change their minds." Another time after we had visited them in our large motorhome and the next year dropped by in our much smaller pop-up camper on a Dodge Ramcharger, he said, "Yer darn near afoot this time, ain'tcha?"

To know people like that, people who helped me change to a Western man from an Eastern boy, old friends with whom we corresponded, and on whom my wife and I called from time to time, was a rare privilege. Just last year we visited their graves with one of Al's daughters-in-law in Woodward, Oklahoma. Tears flowed as I thought of these grand Westerners whose friendship always meant so much. There are some people one can never forget; the world should have more of them!

Martha Belle's "Pig Stand"

Martha Belle and The Pig Stand

It stood by itself on an otherwise vacant lot on the south side of the main street of a small dusty town in western Oklahoma. It was The Pig Stand—in large letters painted on the wooden sign on the roof. Four booths along the east side; a row of stools along a counter with the usual machinery of a café along the west side; a small kitchen in the southwest corner; a back door in the southeast corner; the main door in front facing north. This was Martha Belle's hangout: the place where one could buy the best pig sandwich and the coldest bottle of beer in Oklahoma.

Nobody got drunk in Martha Belle's pig stand. The beer was 3.2, but if you drank six or eight bottles of it at 25 cents a throw, the men's room along the middle of the back would take care of the inevitable overflow. A pig sandwich was 25 cents, with mashed potatoes and gravy 40 cents, with coffee or milk another dime. On an evening before going to our boarding house on a low hill on the southeast part of town for dinner, several of us single guys who worked for Shell Oil Company on a seismograph crew dropped in to josh with Martha Belle and swill down a few beers. After all, jug hustlers, drillers' helpers and assistant surveyors on a 26 man crew could get mighty thirsty working in the 1937 dust storms that caked our sweat with red dust in the hot windy spring of this semi-arid land during the Depression.

The character of The Pig Stand was Martha Belle herself. A tall, wide-shouldered, narrow-hipped Amazon with flashing black eyes, raven hair, and long, strong legs, she was a typical Western girl whose reputation for wild fun was the talk of the town. She was married to a state cop who worked out of Oklahoma City, and every several weeks she would leave on a Friday with a girlfriend to go to be with the big burly cop, who always had a partner for Martha Belle's friend. The older lady who cooked at The Pig Stand was some kind of shirt-tail relative, and she would keep The Pig Stand open Friday night and all day Saturday. Being in western Oklahoma and part of the Baptist Bible Belt, The Pig Stand, which usually sold beer, was closed on Sunday.

Martha Belle's escapades included such things as jumping out of airplanes (with a parachute, of course); riding shotgun on a stage coach in the opening ceremonies of the annual Woodward rodeo; and a number of times on a Sunday when she was in town, riding around in a convertible completely nude—a sight to behold. In a bathing suit ready to swim in an earthen cattle tank on Lewis Wright's ranch north of town, she was a knock out, ready for anything, and sometimes it happened on a late Sunday afternoon with some cowboy while the rest of us were drinking beer and cooling off in the tank under the cottonwood trees. Sunday afternoons were usually picnic times, and it was at The Pig Stand that I first met Lewis Wright and his wife, Al.

The cycle of our work on the crew mapping subsurface structure was ten days on, three days off. This arrangement sometimes gave us Sundays off work, so we could go visit Lewis and Al, drive cattle for him on a Sunday morning, moving herds from one sand hill pasture to another. In the afternoon, we had a picnic with beer we brought to the ranch, bought on Saturday night and stowed in ice-filled boxes. Then a swim to cool off. Martha Belle wanted all of us to go skinny dipping, but I was shy and, I suppose, somewhat afraid of this world-wise, hedonistic hoyden.

It was at The Pig Stand that I learned to drink beer. A friend, Ed Rorex, was an accounting graduate of the University of Oklahoma. He couldn't get a job and became a jug hustler—as I did because I couldn't get a job as a graduate geologist during the bleak days of depression in the late 1930s. One evening after ten days of shooting seismic lines down toward Vici (it was a Saturday night), he said, "Let's go to Martha Belle's and drink beer." I told him I didn't care for beer, so, towering over me in a threatening way, he said, "Come on, you're going to learn to drink beer!" Now there's a protocol to drinking alcoholic beverages if you work on a seismograph crew. First he lined up six bottles of beer in front of me and said, "You only lift the bottle on your left hand and drink it first." Obviously I was thirsty as we both sat in a booth in an unforgettable session during which we each drank six bottles of beer.

The reader must realize that six bottles of 3.2 beer do not make for a drunken spree, but they sure make one's bladder fill in a hurry—and for some rather speedy trips to the toilet. Luckily, or unluckily I suppose, there was no hangover, and from then on in my life, plebeian beer was a favorite when I could afford it. It is important to realize that 3.2 beer was all that was allowed in Oklahoma at the time. A few times in the next months this led to a knock on the door of a small house in the northwest part of town. A small, wizened old man, as bald as a billiard ball, would inspect us through a window and, seeing that we weren't cops, would open the door. In a dimly lit room, one bulb hanging from the ceiling, the purchase was made—a pint of bourbon for three dollars, a quart for five—after which we disappeared into the dark night, car lights out, booze in hand. We found out that a minister or two and some sheriffs deputies patronized the same bootlegger!

These parties were rare, for, after all, we were paid only $90 per month and had to live on that for room and board. The ribald jokes and wild sprees were experiences not to be forgotten. A time or two Martha Belle and her friend were with a group of us that included Harland Heaton, the county treasurer, with his girlfriend, and Ernie Banks, a medical student home from medical school in Tennessee during the summer. Al and Lewis Wright also came plus a few cowboys who worked on the Lewis ranch. Toward the end of the summer, our gang decided to hire the country club for a party and dance. We laid in the booze, all illegal, had a combo, got together a feed, and made a night of it. It was a wild one. Ed Rorex disappeared with an Indian girl in my model A Ford for a while. Some guys and gals, well lit, went skinny dipping in the unlit municipal swimming pool. Everyone danced up a storm, and the party went on until three o'clock Sunday morning. Ed Rorex and I closed down the place with a beer, a sort of last hurrah for both of us, as I was to go back to Harvard, and he had gotten a job as an accountant in Oklahoma City.

The summer of 1937 was a growing-up time—my first in the West where I always wanted to be. Many a Sunday or other day off, I drove cattle for Lewis Wright as a weekend cowboy. The pay: a dinner cooked by Al, beer and a swim in their cattle tank. The friendships I made that summer lasted a half century, until the death of two of my best friends. When I went back East to school, I carried along a small bag full of sage brush, five Charlie Russell prints of the Old West, and a whole raft of black and white photographs. These were to carry me through the next three years to 1940, back in the sodden East, so I would never forget my first job in the West, nor the burning ambition to live beyond "Out where the West begins."

The Scout Check

In the days before telemetering, cellular telephones, oil company radio networks, and commercial scouting companies, there was ear-bending or swabbing oil company personnel for well information. In between, the scout check, held weekly by oil and gas companies to trade information on drilling wells, was the cooperative way a company could keep up with wildcat and production wells being drilled in oil provinces. Still used today in some areas, this is an efficient way to report traded scouting data to management in head offices. Wildcat oil tests, so named first in Pennsylvania between 1860 and 1900, are those wells drilled out in the forested hills "among the wildcats." These wells drilled far from known oil and gas production are important to all companies holding wildcat acreage which may prove to be a new oil province or a new formation previously not known to be producing oil or gas. Most companies divide up regions such as California, the Rocky Mountains, the Mid-Continent, the Gulf Coast, and so on into districts with geological, producing and scouting staffs. Within each district the companies in a larger city will subdivide the district into smaller areas for each company scout to report on each week. Rotation of areas among companies is common, so all scouts get thoroughly acquainted with each district.

There are unwritten rules that govern the scout check. Each is run by a bull scout, who may also be replaced. Woe be it should any scout fail to report accurately each week every possible bit of data on each wildcat well in his assigned area of the district. All scouts assemble weekly in a seminar room or large room of a hotel. All their scout books have a page for each well, organized in the same logical sequence. First, reports are reviewed for all to enter data in the correct place; next come data on drilling wells and, finally, well completions. Each scout reports all the data on his own company wells, unless a well is declared a "tight hole." If so he declares that his company will not report data. This leads to the cloak-and-dagger spy work by all other companies to get by hook, crook, payoffs, bribes, and any other possible subterfuge the data on that tight hole. The scout is not required to give out those data at a weekly scout check unless he wishes to curry favor with other companies drilling tight holes. It is in that spying phase of scouting that the most interesting things happen. Many an oil company secretary has been wined and dined by other company scouts, even bribed, for carbons of scout reports in order to get information on a tight hole that is critical to a rival company. This is commercial spying and intrigue as fanciful as that done by the C.I.A. (back in those days the O.S.S.), and it's a great deal of fun!

The scout check in which I served for Shell Oil Company in Amarillo covered the Texas Panhandle and northeastern New Mexico. For 11 months, between April 1941 and February 1942, as scout for Shell, I truly learned the oil business. Our district was large, and we were required to learn all we could about seismic crews working in the area, all wildcat wells, the most important production wells close to Shell's undrilled leases, any gossip about personnel changes, new types of drilling rigs in use, new methods of well completion, acreage plays "out in the goat pasture" far from production, anything that would give our company a leg-up on everyone else.

Our scout check comprised at least eleven companies, and friendships of lasting value developed there. Many are now dead; in fact, only two of us are left. Allow me to sketch this scout check. Harry Britt of Shamrock was with the only truly local company based in Amarillo. He was a young man who smoked cigars, and one day while being taught how to fly in the old Civil Air Patrol just before World War II, his instructor pulled a chandelle, and Harry swallowed the stub of a lit cigar. This caused general hilarity, for we were tough kidders. Larry Oles was a senior geologist for Sinclair, a fine gentleman who built a bitu-adobe home in Amarillo and stayed after he retired. Dave Henderson of Carter Oil Company was a well-known geologist I had met a year before. Joe Lilly of Stanolind was our bull scout, man-about-town who later became a tail-gunner in a B-17 over Europe in World War II and, later still, married the prettiest girl in Texas. Archie Kautz of Cities Service was a man of prodigious memory, near retirement, who knew everything.

Others were Stanley White, a Britisher of Ohio Oil; J.C. Richardson of Gulf Oil, a helpful geologist and jokester; and Jesse Rodgers of Texaco. Ollie Lloyd of Magnolia spent much of his time working on a large Spanish land grant in northeastern New Mexico and later committed suicide there for reasons unknown. Jack Leveritt of Colombian Gas was a comical friend who was secretary-typist for that company owned by the Cabot Carbon Company of Boston. The last, and newest, of this gang was myself with Shell, after I had completed the vaunted Shell training program in exploration and production geology.

Any oil geologist reading that list will be astonished to note that all but one were geologists, a few near retirement. The Panhandle district was active and, at that time, in a super-mature stage of its oil and gas exploration history. It was the smallest of Shell's several Mid-Continent districts, needing one land-man geologist, myself as junior geologist-scout, and a lovely secretary. Our office was in the Radio Building; scout meetings were held each Wednesday in a basement room of the largest hotel in Amarillo. Each week we tried to stick each other for the round of Cokes for the group. Scouting our eight districts took place each Monday. With 11 scouts involved, we had a week off now and again in the rotation. New Mexico was not actively scouted but was checked about once a month by someone of the check, usually ending up in Santa Fe on a Saturday night for drinks and Mexican food. In fact, Harry Britt and I scouted ten drilling wells in northeastern New Mexico on Saturday and Sunday, December 6 and 7, 1941, to come home to Amarillo to find Pearl Harbor had been bombed. The next day, Monday, December 8, on my scouting rounds in Hutchinson County, I heard President Roosevelt speak to Congress on the radio, asking for declaration of war against Japan. From that day on, my own life was completely changed, and by February 1942, I had applied for a direct commission in the U.S. Navy.

Tommy Knockers

In the silence of a mine tunnel deep underground, in long stopes and mine shafts, high in the cool darkness, deep in all mines, there are noises that are heard nowhere else in the world. They seem to be heard more easily if your carbide lamp is blown out by a gust of cold air which comes from where no one knows. Then one hears a faint tapping noise. Is it a miner with a single jack in a neighboring stope swinging his five-pound hammer against drill steel, preparing holes for dynamite? No, there's no one else in the mine but Willie and me. We're mapping a silver mine deep inside Ramshorn Mountain near Bayhorse ghost town in the Salmon River Range of Idaho. We just sat down to finish up some notes on observations made on veins located by Brunton compass and steel tape. We're drawing diagrams of vein widths, intersections and types of ore minerals, ore veins of galena, tetrahedrite, sphalerite, chalcopyrite,and pyrite. "There it is again, Willie. What is that tapping noise?"

Will Michell
Summer 1938

We were students at Harvard sent out here to Idaho to map a mountain full of miles of tunnels, stopes, raises, gopher holes—a mine originally owned by William Randolph Hearst's father, opened in the late 19th century. "There's that noise again!" Tap, tap, tap, then a sound like a board breaking and a dull thud. We move along quickly to the entrance of the tunnel of Utah Boy #5, one of a stacked set of tunnels extending deep into the mountain from creek level to over 2000 feet up toward the top of bald Ramshorn Mountain with snow still lying in deep crevices in July of 1938. Waiting for us is Jimmy Sullivan, a short, roly-poly Boston Irishman who was the mine mill foreman when the mine was a beehive of activity in the 1920s. Then, 30 miners worked here, sending mine carts full of silver ore down to the mill below to be crushed in a ball mill and concentrated under Jimmy's direction. Now, he was the supervisor, and his job was to guard the entire mine-mill property for Jim Salisbury and the Hearst interests—as well as to look after us while we mapped the mine for Harvard professor Donald B. McLaughlin, consultant to the Ramshorn Mining Company.

Then came the stories as only an Irishman or a Welsh miner can tell them. We sat at the entrance to Utah Boy #5 smoking our pipes while Jim told us the sounds were made by mine goblins called Tommy Knockers. The Germans call them "cobalts." Perhaps they were trolls, the little green men who are the ghosts of men killed during the long and colorful history of this fabulous mine which had produced the millions of dollars worth of silver, zinc, lead, and some gold that helped finance the Hearst Publishing Company. As we listened to Jim's stories grandly told with vivid gesticulations, he mentioned that we would hear the Tommy Knockers all summer long underground, wherever we worked. "Do not be afraid," he said. "They're harmless unless they begin to tap very near you. That's the signal to move out of that part of the mine, because sometimes they tap a bit too hard in the wrong place, and a crushing rock fall will burst through lagging and fill the tunnel." We began to realize that these little fellows were really quite friendly. As long as the tapping was far away, continuous and in several parts of the mine at once, there was no danger. Jimmy, with a puckish smile and an Irishman's storytelling glint in his eye, would exaggerate wildly as he watched to see if we believed him. We could tell he believed in leprechauns.

Jimmy Sullivan was our guide, o ur confidant and our protector through that whole adventurous summer. He was colorful in his language; his exclamations of wonder included "be'Jesus." If he wanted to dress somebody down, he called him "a lousebounder." He warned us: "Beware of the 'dang-ger-ouzes.'" He missed a turn one night on his way to Challis in his car and ended up in the middle of an alfalfa field. Having burst through a fence and into a haystack, he said, "And you know, they reckon time here in Idaho from the date I had that accident." His life is another story. He once played baseball for the infamous Chicago Black Sox! Jimmy died in 1961 and will never be forgotten as long as there's a person alive who met this jolly little Boston Irishman.

Every society has its little people who can do marvelous things, like the "menehunes" of Kawai who build vast breakwaters overnight. The "bunyips," the little people of billabongs in Australia, are of two types: one's a jolly little fellow who guards waterholes, the other, the "bekka," a fearsome, large, hairy creature with claws who lives in the Murray river and means to do harm to everyone. The Maoris of New Zealand have the "taniwha," a mythical bunyip that is the most fearsome king of the forest.

So watch for the little people; they're everywhere, in all countries, subjects of many tales in all societies!

Mr. and Mrs. Jim Sullivan and me
at Ramshorn Mine

Part Five

BUNDLES OF WORDS AND THE HUMAN EQUATION

Books
Writing and the Author
Thieves of Time
The Editor and the Critic
Who Says Nostalgia is Bad?
The World of Men
Always One Left Over
The Abominable "No" Man
The Jolly Liar
The Many Faces of Dishonesty
A Look at Evil
The Negative Equation
Ignorance and Stupidity
Stubborn Versus Tenacious
Rule by Ruckus
Smart Alecks
Achievement
A Life to Live
Disappointments
The Positive Equation
Minimum Lives
Failure
Great Thrills
The Equal and the Unequal
Something to Look Forward To
Bureaucracy and Progress
Why is it, and How Come?
Regularity
Work and Achievement
The Study
The Pesky Telephone
The Piling Theory of Filing
The Pack Rat
The Inveterate Collector

Books

There is something magical about books. To think that marks on paper or stone can convey great thoughts is thrilling and almost unbelievable. The first writing was done with line drawings of objects which led to cuneiform marks and hieroglyphics, complex picture symbols that conveyed ideas of the ancient world. The cave drawings— early forms of writing—of Alta Mira and Lascaux tell of animals the Neanderthal and Cro-Magnon knew. The Anasazi of the American Southwest carved or painted astronomic symbols, arrows as though to show directions, and sun rays to show their knowledge of the sun as a god. Alphabets came into existence—Greek, Slavic, Arabic—and abstruse concepts could be detailed to an astonishing degree. As humans developed their facile hands and time was made available from grubbing for food and shelter, the human mind sought to relay ideas and information. Just as it is said of a country of the blind where a man with one eye is king, so the ability to read and write opened a whole world of knowledge to those who would learn to communicate more precisely.

The relative difficulties of speaking, reading, and writing suggest that speaking developed early. One need only hear an African bushman's squeaking, clicking and tongue-slapping language to realize that talking takes many forms. To read what someone has written is a long step beyond ordinary conversation. To write so others may learn is the highest development of language and the last to occur, whether by a race or by an individual. Without some form of reproducing marks, lasting concepts and the history of a people were dependent on verbal passages from one generation to the next. When papyrus was invented, laborious cuneiform— wedge-shaped marks pressed into clay—was superseded by writing as a means of communication. Moving from labor-intensive hand-scribing; for which we are so very much indebted to the medieval church; to modern methods of print reproduction—printing presses, typewriters, copy machines, on into computer word processing—the world is now flooded with paper. Throwaway newspapers and magazines are reproduced on microfilm and microfiche or stored on tapes and in electronic storage books. The wonder of it all is the book, one of the greatest of inventions!

Perhaps no person has ever so neatly described the resource of books and their greatness as voices out of the past as did that classical scholar Gilbert Highet:

> *"These are not books, lumps of lifeless paper,but minds alive on the shelves. From each of them goes out its own voice, as inaudible as the streams of sound conveyed day and night byelectric waves beyond the range of our physicalhearing; and just as the touch of a button on our set will fill the room with music, so by taking down one of these volumes and opening it,one can call into range the far distant voice intime and space, and hear it speaking to us, mindto mind, heart to heart."*

Ray Bradley said, "Without libraries, what have we? We have no past and no future." Buffalo Bill Cody, with very little formal education, struck a blow for individualism in writing when he was chided for lack of capitalization and punctuation in his first efforts at writing about his life. His retort to publishers was "Life is too short to make big letters when small ones will do; and as for punctuation, if my readers don't know enough to take their breath without these little marks, they'll have to lose it, that's all."

What then is the true essence of books? Many are not worth the paper they are written on, and of several hundreds of thousands written each year, only some 50,000 are published. Many, in fact, are financed by the authors themselves. The magic lies in holding such a store of knowledge in one's hands anytime the notion strikes one. Who knows how many grand and earth-shaking ideas have been drowned by editors and critics in publishing houses who refuse to publish, ideas doomed never to see the light of day in a book?

He who loves not books and reading is poor indeed. You open the world when you open a book! It takes effort to read and even greater effort to write so that others may read. On a street in Albuquerque one day I was asked by a black man where a certain building was that he wanted to visit. I told him it was at the corner of Gold and Fourth Street, and if he would read the street signs going in the direction I pointed, he'd find the building. His pained comment: "Are you kidding? I can't read!" When I was in the Canadian Arctic performing Naval charting duties on an Army transport sailing schooner manned by Newfoundlanders, I was asked by one crew member to read his letters from home. Thinking he needed glasses, for his distant vision was fantastic, I found that he really couldn't read. In fact, he didn't want to learn! Amazingly he was a fascinating conversationalist.

We know that talk dies soon and only the printed word lives on. If one wants ideas to live on to inspire the hearts of those who follow, write them down. Jimmy Cagney, the movie actor, wrote a book titled *The Gallant Hours*, a solid story of his adventure-filled life. In that book he said, among many other wise statements, "It's better to tell the truth than lie. It's better to know than to be ignorant. It's better to be free than to be a slave." Writing for the future is a freeing experience; reading cuts the bonds of ignorant slavery. A writer—perforce looking forward to his posterity while writing of his experiences in the labile milieu in which we live—must surely have looked back at his ancestry by reading as well as listening to family anecdotes of things long past. Be early forgotten, but let the written word be remembered. Nowhere is this as apropos as in a book, for it is usual for some of us to remember the title of a worthy book and forget the author's name. But it is libraries that are the magical places. The American Library Association said in a promotional poster of long ago, "Libraries will get you through times of no money better than money will get you through times of no library."

There can be no doubt that ego is the major reason books are written. Why can anyone presume to think that what he writes will be of interest to others? I am always amazed to hear people—especially retired professors or scientists, or in fact anyone—say, "I love to write books and I do it so my grandchildren will remember me." Or, "I write books for fun. Whether it gets published or not is immaterial."

It's not surprising that most grandchildren and children seldom read books written by grandparents and parents. What spurs one on to write a book may be a combination of ego, money, pressure from a publisher, or some deep well-spring of missionary zeal that can be satisfied in no other way. An author I know published a book that was a blockbuster and then wrote ten others no one ever can remember. Is it true, as one wag said, "Everyone has one good book in his head. How to get it out is the problem"? Yes, the world of books is a dream world, and nothing will ever replace the thrill of reading a good book—or writing one!

Ready to go aboard the freighter *Ivybank*
1991

Pre-Etruscan "writing."
Read from right to left

Easter Island "writing"

Writing and the Author

Nate Buntline, the writer, said, "Once I strike a good title, I consider a book about half done; I push ahead as fast as I can write, never making an alteration or correction." Few of us who write are that capable, yet it is an ability for which to strive. A writer I know says he could write in the middle of a street with a parade going around him, on a busy sidewalk filled with talking people, in the middle of an auditorium during a concert, or in a basketball arena with 10,000 shouting fans. Such would take tremendous concentration. One thing is certain, as Robert Roark said, "No writer has friends, only interruptions." In 1682, John Sheffield wrote in *Essays on Poetry*, "Of all the arts in which the wise excel, nature's chief masterpiece is writing well." Samuel Johnson, who had many things to say about many situations, once wrote, "Toil, envy and want of patrons are the jail of a writer's life." If that is the case, why does almost every educated person strive to write?

Jerome Charyn maintained that "The very act of writing is only a mute's revenge in a talkative world." To do an affabrous job of writing takes time, concentration, more than a little ability, and an ego three yards wide. But there is another characteristic common to writers, as Graham Greene maintained, "There is a splinter of ice in the heart of a writer." I know a young writer who hates all other writers—and says so! An author, when once asked what he was doing, said, "I'm writing a book," whereupon the questioner said, "I'm not either!" Professor Alexander Bickel believed, "All writing is an experiment." A writer soon recognizes good writing in others, jealous as he may be, but one thing is certain: It's only a genius who has the ability in writing to reduce the complicated to the simple. Hemingway maintained that good writing should be done with a minimum of adjectives.

Critics—most of them unable to write well themselves— join publishers—who seldom read well—in knocking down writers simply because too many people want to be writers. Senator Alan Simpson believes, "A critic is a product of creativity not his own!" In the book by John Boswell, *The Awful Truth About Publishing*, are hints as to why so few books are published each year of the nearly half million that are written. The author didn't say that—I did. Dr. Kirk Bryan—a very able writer in the fields of geology, geography and archaeology and a kindly critic of doctoral dissertations at Harvard—gave a general formula for good writing: There should be few big words. There is an optimum number of words in a good sentence; there is an optimum number of sentences in a paragraph. As an aside he would say, "Keep paragraphs short. One a page long is ridiculous." He would go on to say that there is an optimum number of paragraphs to a section and, similarly, for the number of sections to a chapter. Also too many chapters in a book make the book boring. How to determine the ideal number of words in a sentence, sentences in a paragraph, paragraphs on a page or in a section, and so on? The answer: by what sells!

Samuel Johnson made mention that "No man except a blockhead ever wrote except for money!" To further discourage would-be writers, Samuel Johnson made many other astute observations. He might have added discouragement by an overly demanding wife, for many writers seem to require several wives before finding one who will leave him to that most solitary of all efforts—writing! And again, if it sells it's obvious people are reading. If it's drudgery to read, no one will buy. There are two stumbling blocks between an author and success: an editor who will not accept a piece for publication and a critic who blackballs a book or article after publication. The rejection slip is the norm, and the story is legendary about books turned down by major publishing houses only to become major successes when published by lesser publishers or even self-financed by the author. Excellent editors read rapidly, scan chapter headings, begin on many different pages at random, and must have gotten up on the right side of the bed before examining a typescript. They must have originality themselves and recognize creativity immediately. Those jewels in the publishing field are rare, but they do exist in the better publishing houses, so the answer is to submit, submit, submit, and don't give up.

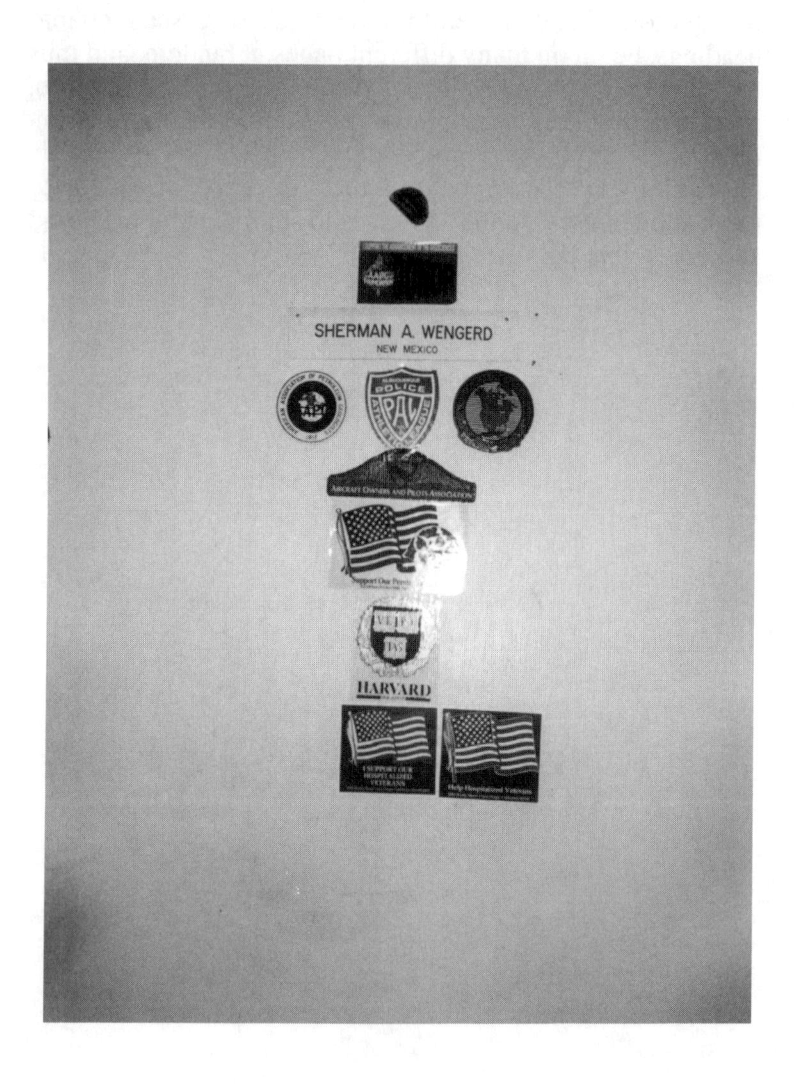

My office door at home

Thieves of Time

If it is true, as some writers maintain, that a writer has no friends, only interruptions, then it behooves anyone who creates to examine the elements that rob us of our time. As a person who enjoys people, it is difficult to shut myself off from the world in order to write. Anything less than a two-hour stretch, with no interruptions, hardly allows the thought juices to germinate and flow from brain to hand and by left hand via pen or pencil to paper. Many writers prefer the typewriter or the word processor and are agile with those intermediate machines. I do well just to make the pen do what my mind wants it to do. I am dependent on my computer-oriented wife and daughter to make sense of my scribbling, interlineations, non-sequiters, and dangling phrases.

But interruptions by anyone, especially friends who telephone or knock on our door at inopportune times, lead me to quote that old trite statement, "With friends like that, who needs enemies." My father maintained that friends—and he had legions of them, as did my mother—not only take time but also cost money. I have found this to be particularly true of friends who are ill, have lost a spouse, are deeply in debt, or just need to talk and tell me their problems.

I write in an office in a building separate from our home—a quiet room filled with books, maps, encyclopedia, and NO telephone. I've threatened to name this study "The Mountains" so my wife can tell telephone callers, "Oh, he's in the mountains. May I tell him you called? Give me your number, and I'll have him call this afternoon or evening." This dodge works fine. Exceptions do occur, as there are certain true friends to whom one just MUST talk when they call. We have for years posted glaring signs on our locked gate and our wooden fence at the street-side of our home declaring BEWARE OF DOG. We don't even own a dog anymore. This subterfuge works, and we are never bothered by off-the-street callers, peddlers, salesmen, and even some friends and other interrupters, with one exception: the United Parcel Service. Those cheeky but friendly lads charge right past the dog signs to deliver ordered packages and leave them by the back door with never a fare-thee-well. They know we don't have a dog!

Once everyone knows about all this, I'll label our motorhome "The Mountains" and actually drive it up to our seven acres of wooded land on the back side of the Sandia Mountains where there is no longer electricity and there never has been a telephone. Besides, it's cool, quiet and away from the hurly-burly of our noisy city. Sometimes when I can't think of anything else to write, I describe the cacophony of our city, any city in fact, which most people never hear, as they themselves add to the noise.

There is one interruption which occurs fairly frequently which I hope will never stop: grandchildren, whether four years old or 20 years old. One grandson walks in unannounced and wants to talk about careers, how to invest money, what should he do about college—nevermind that he seldom listens to my advice—so I know he just wants to talk. At times, I'll have a simple construction job I want him to do, perhaps with my help, and the ideas I've been writing about fly the coop, some never to be thought again. Minor losses are these, for in a grandchild lies my immortality. Another grandson charges in, not through the gate but over the gate, to pull riddles on me or to tell me his latest discoveries. He's fifteen and his voice has changed; puberty is striking this super-intelligent lad. Obviously he takes after his grandmother in the analytical department of his brain, and he is a welcome interruption.

A teen-aged granddaughter wants to know if I have some part-time office work or perhaps need someone to clean up my litterbound study—books and papers piled on tables, low and high, on chairs, on top of file cases, behind file cases, on a drafting table, and on shelves in apparent disarray. Heaven forbid that she rearrange the maelstrom that is my study, though she's a great help in cataloging my library and preparing books to send. If she ever learns to type and read my handwriting (fat chance!), I'll have a full-time job for her: spelling my wife, the avid genealogy researcher who has her own multitude of projects going. She types stuff for me now but at a slow pace; maybe it's because I don't pay her! I tried that many times, though, but nothing ever got speeded up!

And then there are the twin granddaughters who live only a block away. Coy, smiling, insistent that I see their latest toys, the gerbils newly born, their latest art work, or their new shoes. They even ask me when we can take them to the zoo or sailing and camping. Ever welcome, such interruptions, for these two 12-year-old sprites are so non-identical one would think they came from totally different families; they are a joy to behold.

When my own children were on the scene, I had a separate office and study: two rooms in a building away from our main home, "Pueblo Buenaventura," which we sold right after I retired from teaching. I mounted the door handle to that building five and a half feet high with the notion that the age of reason is about eight or ten years, when they could understand that I should not be interrupted while I operated as a consulting research geologist—a sideline to my teaching at the university. This worked for a while until they found a bench and appeared at the window insisting I come out and play, or see their latest creations.

Our children—
Anne, Tim, Diana, and Stephanie
1954

One soon learns to work around such interruptions or in the evenings when they are asleep. When writing a final exploration report for an oil company, I oft-times escaped to our trailer-come-cabin in the mountains which had electricity, heaters, a kitchen, and a large work table where I worked two and three days in a row. All that ended with a real interruption. Vandals burned down the cabin!

All in all one can live with interruptions by those thieves of time by adapting to them or avoiding them or both. Mental alacrity is heightened by such, and they make one efficient. Avoiding procrastination—that most pleasant way of postponing completion of projects—is at times much to be desired. Units of time away from analysis of data, contemplation of new concepts, writing, and revising all help to recharge one's batteries, allow one's subconscious to function, and enable each of us to hold forth and suggest solutions to old problems as well as inspire the germination of new ideas. To be forever right and always working makes one a dullard. Isn't it nice that doers can once in a while rest? To do so is an important freedom. There is a Mexican proverb that goes, "How nice it is to do nothing, and then, after doing nothing, to rest." OK, if it's not overdone. I hail the thieves of time as long as they don't steal too much of it.

With my twin granddaughters
1981

Top of my desk at home,
filled with family photos

Editor and the Critic

It was Art Spaulding, an early officer of the American Institute of Professional Geologists, who found what happens to aging editors: "Old editors never die. They just fulminate away." The late Frank Conselman—that biting linguist whose wrath was aimed at editors, though he himself was one of the best—maintained that STET meant "Style transcends editorial tinkering." Some of the prime characteristics of a good editor include the ability to recognize misspellings, bad grammatical constructions and plagiarism. These require a wide acquaintance with the literature, a deft hand at tightening up sentences, and the capacity to be insulted by misplaced letters. One might call this ability cognitive learning or factual learning in depth. When speaking of journalists, what Spiro Agnew said may be applicable to some types of editors and critics: "They're an effete corps of impudent snobs."

There is a general notion among authors that people who are cogent critics and thorough editors cannot themselves be good writers. That may be true of some in fields other than science, but I have found as an author of many scientific articles in geology that the best and most patient editors are also good writers. Of all the critics I've encountered, there have been several who were highly bombastic, given to argumentation ad hominem, stilted in their criticism, and were generally obnoxious. In every case they were people who didn't know enough to criticize in a special field, or had themselves been scooped by another author and thus tried out-and-out deception to keep an article from being printed first. It may be better to submit elsewhere or publish privately. When an editor tells you that you write badly—"I don't believe what you say," "Your documentation is shoddy," "Your facts are suspect," or "Your contribution is unfit for publication"—look behind the comments for reasons which may have little to do with the subject at hand.

Few nationally-elected scientific editors are knowledgeable in all sub-fields. For reasons I never understood, editors will not identify associate editors to an author. The chief editor will blandly tell an author, "It is so that the associate editor can be utterly objective and not be subject to backlash." That is completely incorrect! When I was asked by a national editor to serve as a critic and associate editor, I insisted that I be identified to the authors whose works I critiqued, good, bad, or indifferent. Today many of the authors whose works I turned down initially, backed by the chief editor, are still friends of mine. Why? Because I went to great effort to help every author to rewrite, document additionally, and even research later published works which would bolster his article or prove him wrong.

There is another facet of editing that becomes moot as to content. I believe in the concept that an editor who has NOT done an equivalent amount of scientific analysis in a special sub-field covered by an article should give specific reasons for saying "your facts do not warrant your conclusions." It has always been my contention that an author "should be allowed to hang himself," as Dr. Kirk Bryan used to say, and let the readers prove the author wrong as to conclusions. Most rebuttals by older scientists to conclusions reached by younger scientists are rife with error and, in many cases, indicate shockingly the existence of professional jealousy. Could it be that the senior scientists had not thought of new concepts based on precisely the same set of facts?

In general it is good policy to follow the advice and suggestions of associate editors or, as senior editor, to convince this very helpful critic that the author may be right—that perhaps the author only needs to tighten up his text and recast his conclusions from dogmatic statements to the use of the more diplomatic "possible," "probable" or "I believe." I know of two major scientific articles in geology that would never have been published if a national editor had not literally fought tooth-and-nail against the turn down by several less imaginative associate editors. Those two articles today are recognized as having advanced the field of geology into a revolution in thought about the origin of the earth's great basins of deposition. Knowing this, I often wonder how many earth-shaking concepts have been buried in the round files, the dust bins, and the fireplaces of authors when works have been turned down by unimaginative editors. If the contribution fits the space, the author makes requested changes, and the whole article isn't some cockamamie story without basis, PUBLISH IT, then let the critics blast away at the author. In my past experience, critics may blast away at an editor rather than the author. Critics who do this are the type of individuals who kill the messenger because they don't believe the message or don't agree with it.

In a noteworthy decision, when I was national editor for the American Association of Petroleum Geologists, I was severely criticized for publishing a long article on carbonate classification by a young professor who had worked on the subject for years and refined it in the laboratory. The critics maintained I should have waited until a major foreign company got its article ready because, they said, "Don't major companies generally do better scientific research as teams than a lone individual at a university?" Obviously that's not always true. Lewis H. Lapham, in *Money and Class in America*, is responsible for a quote involving such an attitude toward individual scientists: "Most Americans cannot bring themselves to trust the unaffiliated individual." That criticism came after the carbonate article appeared in our national bulletin. I wish I could publish the letter interchange with that major company which involved the very old basis of "We publish what is acceptable when it is offered and wait for no one—big, little, prestigious, or unknown." To this day no contributions on carbonates by that foreign company have been made widely available in the AAPG Bulletin.

The old adage "publish or perish" does not refer, as so many think, to professors who must publish or they are not advanced in rank or salary at universities. Although that is a mistaken impression in many universities, it was William Morris Davis in 1898 at an International Geologic Congress who said, "Publish or the concept will perish." His reference was direct and correct; if you develop a well thought-out new concept, publish it as soon as possible! Prior right is important; don't be reticent, fear criticism, or be reluctant to fight for the right. The greatest scientific concepts hidden under a bushel are of no avail. As my first professor of geology in college, Dr. Karl Ver Steeg, taught all of his students, "Blow your own horn, or no one will even know who you are."

All of us who have made somewhat of a reputation, whether as authors, editors, critics, or a combination, are indebted to our mentors and predecessors who led the way and recommended that we be considered for important positions in science. Such a one in my life was the brilliant scientist Dr. Frank Conselman, who said, when he relinquished to me the job of national editor of the American Institute of Professional Geologists, "This is my swan song. Here comes the relief crew!"

Who Says Nostalgia is Bad?

A backward look often gives one forward direction, and this nostalgic activity should be most actively pursued in retirement. Why? Because if there is ever a time to look forward, it's when the present isn't cluttered up with efforts to maintain shelter, food and security. In fact, to plan one's life as though you would live forever and live each day as if it were your last means that during the present day you are, in fact, planning activities that keep you alive for tomorrow. This often leads to nostalgia, for how else but through one's experiences can one project into the future? Eat, drink and be merry may be a great idea if you plan to die early, but one thing is certain: a look back gives satisfaction when appraising a life well lived. To think of the many forks in the road of an active, thinking life makes the true philosopher realize his debt to chance. There is a certain balance that results from introspection which, in turn, leads to creativity. A love of learning and deep relaxation typify the mental mobility and the incredible nobility of the intellectual activist. Although there is a challenge to unscheduled living, most people are indeed in a rut and, if they left it, they'd be lost.

I have a number of acquaintances who seem to have become rather dimwitted as they sit groaning about their pasts and how, if they had it to do over again, they would lead more adventuresome lives. Mentally handicapped by long consideration of what might have been, they forget how exciting their lives really were. One of these chaps was a pilot in World War II. Drawn to drink by the thought that he should have become an airline pilot, he moped about telling stories of his exploits over Germany. But did he consider writing about the many dangerous encounters that led to his being an ace? A man can make nostalgia a valuable tool, given the gift of writing and more than a little capacity for exaggeration. In fact, as one gets older, the great and good things that happen get blown up out of proportion and become fiction. And it is fiction that dramatizes nostalgia.

The world is full of events about which to be nostalgic: the time you saw Lindbergh, the first time you rode on a ferris wheel, the burning of the mortgage to your home, the last cigarette you smoked, and so on and so on. If nostalgia makes you feel good, go to it. You might be inspired to think like one friend of mine. He said, "You know how I want to die?" I thought, "What a macabre thought!" He went on: "I want to be shot at the age of 90 by a jealous husband as I climb over his back wall after having made love with his wife." Now there is a futuristic situation brought on by nostalgia!

Once the workaday world is left behind, controlled nostalgia becomes a powerful tool that sustains life if positively used to gain new experience. Lives there a human so dull that reverential hindsight does not provide inspiration? A grandfather tells his grandson about the old days. Of course, the grandson hardly listens if Grandpa is obviously preaching. Yet a number of years later a thought strikes the grandson: "My grandpa told me about that. I'll bet I can use that advice right now." And so if your knowledge of a well-remembered past leads to progress in someone's future, what in life can be better than that?

The World of Men

Every man born of woman—as all men are—is half woman. Assume this to be correct, although it does not take an astute observer of the human condition long to realize that there are as many very effeminate men as there are manly women. Leaving aside any discussion of the philosophical concept that every woman is also half man or that there are numerous homosexuals among us, I seek to consider only the unique capacity that men have for having a jolly time together in many endeavors. Whether it be flying, golf, drinking, sports, womanizing, checkers, writing prose, handball, sailing, chess, carpentry, whatever or whether it involves vocation or avocation, there is a camaraderie among men of similar interests that transcends competition and may even exist because of friendly competition. Some men whom I know—and I associate with groups from all stations of life—are gossips; most are jokers given to ribaldry, outrageous practical jokes and laughter. Exists there a man so dense that he doesn't like a good joke, or even a bad one?

I seek to make no invidious comparisons or contrasts of men's characteristics with those of women, although it takes all the discipline I have to avoid such. To me the world of men is a dynamic one, and the concept of mate, buddy, friend, or close association of one man with another is a bulwark against the world. This is not true of the homosexual relationship, which is never at ease. I make only a few further comments about that most unusual and aberrant association which springs from deep within the gene pool and is difficult to submerge, or in popular parlance, "To keep hidden in a closet." Few homosexual men have lasting friendships within groups of normal men. They are aliens in the world of men. It has nothing to do with being effeminate, as I have an effeminate friend who has raised two large families, in succession of course, and is as tough as nails in a fight. It also has little to do with the common concept of manliness, for I have two very manly friends who are abject cowards in the face of danger. I realize that such anecdotes are not statistically valid, but they are true!

If I have established in the mind of the reader a threshold concept that there are men who look what they are not and other men who are not what they seem to be, then a calm discussion of the world of men is possible. I belong to many groups limited to men of all types: tall, short, strong, weak, skinny, fat, courageous, cowardly, healthy, unhealthy, tom cats, womanizers, shrinking violets, whatever. These include voluntary groups such as service clubs and faculties, among fathers, within scientific societies, in flying clubs or groups of writers, and include almost any group apt to be male-dominated, to the horror of rabid feminists. "Wait a minute," you may argue. "There are women writers, women in scientific societies and in flying clubs." I agree, but I'm considering groups of men who like to be together by choice whether members of larger groups or not.

I also avoid categorically the concept by so many feminists that the world is unfair to women because we use male pronouns, categories with the letters M E N in their names, such as chairmen, human race and so on. One of the most stupid words I have ever seen used in academic circles is the word "chairperson." How can a person be a chair when he or she is a chairman? It is blasphemy to rewrite a language because of the objections of women, and men, who have nothing better to do. A violent backlash is in order—now!

A deep appreciative bond exists among men that I do not see among groups of women, yet, not being a woman makes that statement highly suspect. I am quoting several women whom I admire who tell me that it is a fact. Women with whom I have a platonic friendship tell me of the jealousy they feel because men have such a good time as men. I doubt that jealousy is based on the oft repeated phrase "penis envy."

It MUST be related to the observed easy camaraderie of men who seek no advantage one over the other, men who enjoy each other's company. What more can be said? Scientists tell us that if you can't document something, measure it or duplicate it, it must not be true. Don't you believe it. There is something special within the world of men. Let's keep it that way!

Our sailboat

At the helm

Always One Left Over

It has always seemed strange how many humorous lists of activities I've collected that indicate the onset of aging. Some are given to me by friends. The latest was sent to me by a valued friend, Woods Hinrichs, the gentle giant I seem to have known since I was a boy. He sent it along with a xeroxed letter telling me how much he values my friendship through many years, with a penned note: "Thought you'd get a kick out of this." There are many criteria that tell me years have passed by all too rapidly. I don't feel old—perhaps older, but darn sure not old. These criteria illuminate situations that happen almost every day, as I shuffle papers, use tools, try to put writing sheets and pens away after four or five hours of writing. It is best described as "Always one left over."

The day's work is done, no matter what it is: building a wall, cleaning a boat, putting together a new sawhorse, cleaning up a mess after stripping an old clock, anything that involves several of something that must be stored "so they" can be found again. "A place for everything, and everything in its place," as my mother used to say. I clean up everything and it's all ship shape—maybe it took 15 minutes to do it all—and I've finished putting it all away, ready to watch the late afternoon news or a baseball game on television. Lying right in plain sight, as though it jumped into view from the store of stuff already laboriously put away, is one thing I forgot: a pen, a notebook, a chisel, a sheet of paper, a hammer, anything that is inanimate and shouldn't be there in plain sight mocking efforts to be neat. Why is this? I am convinced that inanimate objects have minds of their own; they are usually obtuse!

Make a list of things to do—places you need to go, things to buy. Write it all down so the items can be numbered in some logical sequence and no time is wasted, and you know what? One of the most important things or plans is forgotten and doesn't get on the list—always one left over! "What the heck was it?" Don't worry. It will come to mind as I go along, but at times it doesn't until I've done what the list says. I'm finished and it pops into mind. Too late for today. Not to worry about forgetfulness. Do it tomorrow. I console myself that it isn't aging but rather the fact that I do myriads of things in any one day. "Be thankful you get as many things done as you do," I tell myself. Who, me wrong? Nah! Tomorrow I'll knock 'em dead.

There are many lists extant about aging. This is a diatribe about the perfidy of inanimate objects and the vagueness of a busy mind; one may mention many things that did not happen to me, and I feel better. I've talked to my grown children about these human frailties, and even in their youth, they tell me it happens to them as well! It does mollify me, but it's irksome to do something all over again, to put away one object leftover, to clean out a leftover brush which you used to paint a picture frame but have already put the paint thinner and tin can away. Why do these things happen? It's got to be some kind of perversity, wickedness or obstinate characteristic in things. Yet I realize it's my fault. Next time it won't happen, and I go blithely along. Forget it. Life's too short to worry. To heck with it! Ever onward.

I counsel with a group of guys—the P.O.E.T.S.—over a beer at the Elks Lodge every Friday for lunch. Their retort: "Join the crowd. It happens all the time. Think nothing of it. Just happened to me this morning," and then they trot forth a litany of small failures of the human senses and the mind. I feel better, but I wonder if such might not be a frailty that comes on with the years. I don't fear aging, and I'm not afraid to die. I think of a younger friend of mine about to undergo heart bypass. He asks the nurse, "If I don't make it, will it hurt?" Does it really hurt to get older and more forgetful? Not if you don't let it bother you.

So in the scheme of things and activities that fill a day

with overflowing achievement, life is good. It's all great fun, and then I wonder what would happen if I record all my thoughts in the sequence in which they happen for one whole day? Would the listing itself spoil the experiment? I'm convinced that some important thoughts would slip off into that ungraspable void of forgetfulness and simply not be recorded. Always one, or more, left over!

Me (second from the right)
with seven of the P.O.E.T.S.

Showing my “Helirangs” to a friend
1981

The Abominable "No" Man

The world has in it many nice people. Most will agree with you if you are sincere and don't bore them to tears with stories about your difficulties or your operations. There is one group of nincompoops who ruin all conversation and seem to have no real thoughts of their own, except contempt in response to what other people say. They are the abominable "no" men. I limit this to men because women, who have the same tendency to challenge the ideas or facts of others, are better called the "yes, buts." Both are strange breeds of contentious human.

Every abominable "no" man I've ever met is at best an angry man. Most are sexually deprived, presumably because they can't get it up anymore. Age is immaterial; many are healthy but have lost their zip. They're easily insulted, seldom laugh and, if they smile, it's a smirk of derision. A few of them are naïve; some are intelligent and make sure you know it. They respond to a question with another question, seldom commit themselves to a positive statement, are fearful of rejection, yet are domineering. If they appreciate women and are able, they go for one night stands, but platonic friendships with women are alien to their psyches. If challenged in their negative statements, their retort is inevitably, "Don't you trust me?" or "You think I'm lying?" or "What do you mean by that?" when they know perfectly well what you've meant when you contradicted their deprecations of what you've just said.

They tip their hand so readily when they say, "No, but this is the truth and you'd better believe it." Like an outright liar, they are not dangerous, just disgusting. Why do we put up with them? How do you get away from a boss like that when you actually enjoy your work? Find a new job? How do you avoid a neighbor who's always saying,"No!"? He borrows your tools and misplaces or breaks them. You find them hanging in his work shop. "Oh, I forgot that was yours." Or, "I just bought that," and there, in plain sight, are your initials on the handle. They are the types who borrow money and forget to pay it back or invite themselves to drink your beer and never have any on hand after you help them jack up their boat trailers for the winter. They think it's great fun when their dog chases your cat up a tree. "Well, if you'd keep that darn cat in the house, my dog wouldn't chase it." When their mutt digs up your newly planted garden, they retort, "Must be someone else's dog. Mine is so well behaved." Also, their kids never do anything wrong.

People like that are easy to detest. Nothing will mollify such a misanthrope. If you commiserate with him about his son's suspension from school for a week for smoking in the boys' toilet room, the "no" man raves and rants about how unfair teachers are. When such a parent shows up at PTA, he challenges everyone else's ideas with aspersions as to other people's intelligence or background. Roundly turned down when stupid suggestions he makes are brought to a vote, he storms out of meetings. Luckily, this makes everyone there happy.

What is the answer to such a personality? One always thinks too late of a *bon mot* rejoinder, the tart comeback that makes him look like the heel he is. But if you do fire back in front of others, it passes right over his heads. If psychologists really want to do some good for the world, they'd study such idiosyncrats, and let us know, first, how to correct them, and, second, how to successfully avoid them!

The Jolly Liar

In a local newspaper in Wilton, Connecticut, author Paul Theroux chanced to see a quotation of his he had concocted in a book review written years before: "I never knew a snob who was not also a darned liar." This rang a bell in my own experience about snobs, so let me tell you that I never knew a jolly liar who was a snob! The jolly liar may not be that at all; he may only be a storyteller who exaggerates. Certainly such a conversationalist is preferable to the officious person who must be absolutely exact on everything and always right—the "yes, buts" of our society.

My wife and I have a friend who is a "yes, but" who tells me frankly she doesn't ever know whether to believe anything I tell her. Even facts! This not because I've ever told her a lie, not even a white lie, deliberately, but in conversation my voice sounds as though I'm exaggerating, which may occur inadvertently at times. Conversation, to be interesting, requires that wisdom be exchanged, not in great streams of words that mean nothing. I may be wrong, but people who talk a lot and say nothing must, by the very volume of words, be lying more than a little. Perhaps if they are jolly about it, one could enjoy their nonsensical jabbering if they at least add notes of humor, even if misplaced, and a bit of lying.

If one hears a person deriding behind his back the motives of another or how he or she looks, talks, and acts, you can be sure the derider will do the same about you. It's like that honor bit. If they talk about how honest they are, listen carefully but watch out. The psychological ploy, the ulterior motive, the jealous attack of one person on another is easily identified and certainly not jolly. A person who is an outright liar—and I have some acquaintances who are just that—is very interesting to trap with facts. If he is a purveyor of bad judgments, one can forgive mistakes, but outright lies may be interesting measures of a man's character, if delivered in a jolly way.

We are all psychically dependent on each other. All conversation should be fulfilling a need to be loved and looked up to. If the conversationalist has a pleasing voice and a happy smile as he makes you feel superior, that person is worthy whether or not he tells you a lie now and again. If it is a friend and you disagree with what he has said in a diplomatic or humorous way, he may only momentarily be taken aback. "Bullshit" has stopped many a jolly liar in his tracks and made him laugh heartily. If you use that stopper with an acquaintance, be sure to duck in case he comes out swinging!

Many people are utterly dependent on good conversation. "No man is an island," writes John Donne. We are all interdependent. But if you run into a person who just listens and says little, perhaps he doesn't hear well, or he may be a bit senile. Such quietude led comedian Red Skelton to observe that "Just because the tea kettle don't whistle no more, ain't sayin' there ain't nothin' cookin' inside." Some people are so serious that they must realize that one can't learn by talking. To draw out such a person, a jolly lie will make him rise to the bait with an exclamation of disagreement, and a two-way discussion could begin. In another context I've mentioned the dictum that one can double the quality of bull by cutting the quantity in half.

There is an aristocracy of excellence among conversationalists, and those who are dull consider a good conversationalist, liar or not, to be an elitist. What they don't understand, they are quick to doubt. Some people have so much fear of getting egg on their faces that they say nothing. Might as well go out and talk to a fence post. I almost said a tree, but then I remembered a petroleum engineer out on a wildcat well in the desert of Utah who spent time talking to a large piñon tree, convinced that the tree could hear him and was happy for his company. Each jolly liar to his own peculiar beliefs!

I have never met a good conversationalist whom I found uninteresting, even if I didn't wholly agree. A little exaggeration and a lot of humor are good for the soul. So long as words mean the same between you and me, a happy rapport can exist. Good conversation defeats boredom; only would the incessant talker who interrupts and won't listen learn that simple fact. Every person who is immediately likable on first meeting is invariably an interesting conversationalist. Such are to be blessed as friends. If a palpable enmity is obvious, it's best to say nothing, for it seems that such people harbor ulterior motives which can be harmful. Amity is the one beautiful characteristic of an interesting person, jolly liar or not.

My sister Carol Ann
with Mom and Dad
1934

The Many Faces of Dishonesty

There are few books as compelling as the Bible in the views of ancient Hebrews, concerning the many forms of dishonesty among humans. I am reminded of Psalm 34:13, 21: "Keep thy tongue from evil and thy lips from speaking guile. . . And they that hate the righteous shall be desolate." The many forms of dishonesty have spawned myriads of words ranging from insincerity to murder. The frightful thing is that these words do not make a linear series or a stochastic sequence, but a broad area involving intention. Across a line of insincerity or murder lies a path that ranges from sophistry to malfeasance.

Anne Morrow Lindbergh thought, "The most exhausting thing in life is being insincere." To me insincerity is a mild form of dishonesty, a misleading human characteristic that typifies a fallacious mind which distrusts truth. Perhaps a third dimension of the tapestry of untruth is revealed by a line ranging from exaggeration to outright deceit. When one adds the fourth dimension of time, the great globular area of dishonesty is able to house a volume of dishonesty in which gradations occur in the same person as his life goes on. This metaphor does not just involve synonyms for dishonesty, but it also involves grades of criminality within the vast ranges of human intercourse. Some untruth is malfeasance gradating to active malevolence; from harmful words to actual physical harm; from simple white lies and ill-will through active lying to treachery; from slanted facts to mendacity and calumny, resulting in harmful and intentional perfidy; from simple, self-serving corruption to all-embracing evil by an entire individual tribe, state or nation. Hypocrisy fits in there somewhere!

There are numerous malcontents and wastrels in every society who cut corners to gain advantage and destroy character through slander and defamation. Among writers, stealing another's words and phrases is plagiarism. I rise to the defense of writers who use the quotations of the famous, the infamous and even the unknown among us to illustrate points and philosophical appraisals to the human scene—for quotes are the stuff of truth as those quoted saw it, and they belong to the public domain. Quotations found in lectures, newspapers, magazines, and, yes, even books are not plagiarized. If the author is named, a service is done much more valuable than hiding behind "anon." I once used a quotation by a man named Quattlebaum involving the pursuit of happiness as being off limits for backing by government funds. A liberal friend of mine thought that was a low blow because he couldn't find the quote in *Bartlett's Quotations*! Frank Colosy had such a person in mind when he said, "A lopsided man runs fastest along the sidehill of success."

Few of the above terms for forms of dishonesty are actual synonyms. I'm reminded of a radio broadcast by Paul Harvey in which he said, "A synonym is the word you use when you can't spell the other word"—or when you can't pronounce it! *Roget's Thesaurus*, valuable as it is, doesn't have many of the synonyms in use today because our language concerning dishonesty continues to grow and become burdened with more and more words as dishonesty becomes more pervasive in our society. Any book on synonyms, quotes and sayings is obsolete by the time it is printed! Any untruth can be magnified to become mayhem, and we have many nouns to label wastrels and similar persons. Note the variable grades of prevarication, equivocation, fabrication, distortion, misstatement, and disassembly. Can these be put into some logical, gradual order? Not likely! About as possible as ordering cows, snakes, fish, turtles, frogs, oranges and bananas into some logical sequence depicting their value to human diet.

Dashiell Hammett once observed that "The cheaper the crook, the gaudier the patter." Aside from physical mayhem, the human tongue is the most dangerous tool of dishonesty. Unfortunately, criticism is often based not on truth but on jealousy, and such criticism is dishonest. A wordy denial of known facts labels the person who denies the truth as unreliable, a person unworthy of further consideration. The "yes, but" who always contradicts every statement is usually a liar. It is best to avoid such persons. Abraham Lincoln, that paragon of honesty, was shrewd in his observation that "No man has a memory long enough to be a successful liar." Mark Twain suggested, "When in doubt, tell the truth."

Eternal vigilance and turning a searing light on wrongdoing can root out dishonesty. At the risk of libel suits, blame should be placed publicly. Ethics committees seldom find fault with congressional misdemeanor. "The fox guarding the henhouse" syndrome is a self-serving piety. The basic corrections may come from within, but failing that, they may come through vengeful retribution. There are several rather famous observers who have made statements about overly defending oneself from thievery—a heinous dishonesty. Dr. Samuel Johnson said, "And while they speak of their virtue, why, sir, when he leaves our house, let us count our spoons." At another time he made this observation: "If he does really think that there is no distinction between virtue and vice, when he leaves our home..." Mark Twain put it this way: "And while he speaks of his honor, let us count our silver." It is difficult to know if Mark Twain copied the idea from Samuel Johnson or thought of that truism himself. In fact, Ralph Waldo Emerson has been quoted with the same thought! Not to mind, it's a truth. Be watchful of people who tell you how honest they are, and you must distrust loquacious liars who protest too much.

If a vice hurts only the perpetrator, a crook shoots himself in the foot. If a bank thief wrecks his own car in a get-away, an investor loses money in a scheme that backfires, a robber is stabbed by another thief to gain all the spoils, so much the better. It's too bad they weren't eliminated altogether.

As long as there are people, there will be dishonesty.

It behooves us to try the following: straighten them out with truth, enlighten them if possible, publicize their perfidy, utilize retribution unbeknownst to the person, or avoid them altogether. They are unworthy leeches on society. Eventually their lies will hang them, but sometimes not soon enough!

A Look at Evil

There exist in the world of men and women individuals who are abnormal in their attitudes toward the rest of us. It is difficult to understand the insane, but these abnormal individuals are not stupid, nor do they lack the intelligence to function efficiently. Some may consider them insane simply because they view life and other people as pawns to do their bidding. Among men it is the warlock, among women the witch. Granted that all of us have within us a trace of witchery inherited from days of long ago, it is difficult to understand why one would cause deliberate personal harm to oneself, such as masochism or sadism, the deliberate and calculated cruelty to cause destruction beyond all reason.

A milder form of evil involves the scofflaw who actually infects us all at one time or another, as we see bad laws enforced or good laws enforced badly. These aberrations in human character may lead to irreparable harm, stunted ambition, a railing at fate, disappointments toward living well, and discomfort uncalled for as one human creates inhumanity toward another. A highly educated friend of mine in Dallas declares that an old woman he sees in a cafe some days is a witch. He is certain she casts spells, that evil things happen when she is around, that even to look her in the face tempts her evil eye to cast bad luck within the hour.

Who has not met or seen a warlock or witch, a man or woman whose very existence seems to blot out all the good around us? If a warlock, it is a man whose predilection to evil is so great that harm comes to all who associate with him. He may be a Jonah on a ship, a dean at a university, an arbiter dismembering an oil company, a spell-binding preacher leading people to death on religious grounds—as at Jonestown—or someone you know who is trapped within an attitude that the world is against him and uses every dirty trick in the book against friend and foe alike. Such fools we mortals be!

To what in the gnarled mind of such a person can we attribute deleterious actions taken without apparent cause? How can we protect ourselves against such evil individuals? Men and women to whom normal social intercourse is completely alien, who destroy for the joy of mayhem, are legion in literature, even to the point of self-destruction. There are those who prevaricate simply because lying is more fun than telling the truth. Such an evil man was Errol Flynn, who admitted as much. The Roman Emperor, Caligula, was another. One could also include Judas Iscariot. The concept of the Devil, a character blamed for all kinds of mischief which we perpetuate ourselves, is a favorite and the supreme warlock. Every society has stories of sell-outs to the Devil to gain immediate advantage: the Faustian Syndrome.

From the revenge-seeker to the murderer, there is a complex gradation of evil. J.W. Drakeford said, "The id is a primitive force, the ego the decision-making self, and the super ego or greater I, the conscience." In 1936, Harper and Row published a book by Anton Bolson titled *The Patient's World* in which is the phrase "Mental illness is really a disorganization of the patient's world." When conscience is missing, evil takes over. During Puritan times it was believed that unbalanced people were witches, and even today in some cases, as in the Bible, we think of the insane as being possessed of evil spirits or the Devil. Though warlockery and witchery are true aberrations of a twisted mind, those people are too shrewd to be considered insane. Try to understand the criminal mind without falling into the incorrect morass of "the environment made him so"—the explanation so dear to super liberals who forgive evil and castigate law and order, blaming crime on family backgrounds.

The two worst forms of evil involve terrorists and the actions of the Japanese toward prisoners during World War II. How can such ever be excused? Roguery and rascality have no place in the civilized world, for their evil afflicts great numbers of people even today. Religionists tell us that "to know all is to forgive all. Pray for those who make our lives miserable. We must become imitators of God." Treat evil with kindness and you're dead. The Old Testament's "an eye for an eye" illustrates that the ancient Hebrews knew how to handle evil. It lowers the righteous not one whit to treat evil with immediate retribution. Vindictiveness has an important place in the fight against crime, for the evil understand only power to dominate. To do less is to lose one's soul in a dangerous world.

Boarding a freighter

The Negative Equation

In any large group of strangers thrown together on a ship for a long voyage, there will be from one to six people whom one immediately likes and two to four who must be avoided at all costs. Groups of 20 to 30 people will show this ratio no matter what their wealth, station in life, sex, or diversity of interests. On freighters there are six to twelve passengers on most lengthy voyages; of those, one to two—the two usually being a married or unmarried couple— will have the personality of a rotten persimmon or a sour pickle. Loud, drunken, condescending, not particularly bright, and usually well-to-do, they are jealous of doers and they create frictions. If you are unable to avoid such misanthropes, your days will be spoiled. They obviously are acerbic because they have bile for blood!

The interesting part about such people is that they can't be recognized as being a single body type. Everyone can readily peg the Adonis type who primps and preens and thinks himself God's gift to humans or a woman who just knows she is God's gift to men. A spoiled woman stands out like the sore thumb that she is, especially if her doting husband tells everyone within earshot how wonderful she is. I'm also suspicious of women who loudly brag about their husbands in public; best to leave such encomiums for the bedroom. And then there are the characters of high pride who, as red necks say, "Haven't a pot to piss in or a window to throw it out of." I've never met a person of good humor who was a negativist uninterested in other people. If time wounds all heels, why doesn't it do a more thorough job?

It has often amazed me that these negativists in the equation of life do not realize their unpopularity, and they have so little sense that people, normally in close association, are trying to avoid them. They are not enemies—no hatred need develop—yet they are disliked, and conversation with them is purposely kept to a minimum. On one outback safari in northern Australia, there was a woman—divorced by her Norwegian ship captain husband, moderately pretty, talkative, desirous of being waited on by everyone—who became so incensed that the married men in the group were avoiding her that she blew up in violent anger. During the whole 18 day safari, she never was able to strike up a normal conversation. Once on a voyage in the Galapagos Islands on a sailboat with 11 people plus a crew, a rather pretty but completely spoiled woman made life miserable for others and gave our guide fits with her demands. Later we found she had not paid her fare to organizers of the expedition, could not be found after our return, leaving our guide, who had recommended she be taken along, to pay her fare. He'd trusted her because her divorced husband was the guide's friend.

There must be a streak of dishonesty in people who are perverse in personal relations with strangers. There is nothing easier than to be friendly, but rebuffs by snide remarks, obvious jealousy, lack of cooperation in group activity, anger at little problems, and pushiness all point to an imbalance, a flaw only a few steps from insanity. Of particular note is the type of person who is always right, loudly proclaims how you are wrong, becomes abusive while snockered with alcohol, must have the best seat in the tour bus or the best cabin on the ship, or always sits at the captain's table for dinner. Best to avoid such persons!

Oscar Wilde, who said quite a number of unbelievable things, did make several bright comments, such as, "There is no sin except stupidity." But it isn't only the stupid who are unbelievably difficult. So are many selfish intelligent people. Dean Alfang's statement about choosing to be different doesn't permit one to be obnoxious. The things that drive most people to excellence are pride, approbation, acceptance, sex, search for comfort and security, ego, and enthusiasm. With moderately good health, anyone can be friendly. If they can't be friendly, it might be due to ill health, bad hearing or stupidity. Though allowances must be made for health difficulties, an intelligent person in poor health need not be unfriendly. Ben Franklin maintained that "A fool and a rogue are but a minute apart." Hence, one must conclude that perverse people are mainly fools, rogues or embittered iconoclasts.

Time spent with persons of ill mien is another waste. They will not rise to good conversation nor contribute memorable thoughts. Some talk only of inconsequential things, rail at people, seek only their own satisfactions, and get in the way. In any philosophical discussion—which few such paragons of stupidity can understand—they always easily doubt a new opinion or idea rather than take the time or make the effort to comprehend it. One may wonder if such people really like themselves, for it is a fact they like no one else, and he who likes no one can never be liked himself. In unsuccessful attempts to be friendly to selfish, jealous people one meets in group situations, one must remember Zorba the Greek's postulate: "On the door of a dead man, one can knock forever." Patience is a virtue only when it is productive, and I've found conversations with negativists to be wholly unproductive, except for this: one learns never to act like they do. Elbert Hubbard said, "Fences are made for those who cannot fly"—a clear invitation for misanthropes to stay at home because they lack the imagination to be agreeable among strangers. They SHOULD be fenced in!

Another characteristic of humans on the negative side of the equation is that they seldom apologize for their misdeeds. False gossip is their stock-in-trade. Denigration of everyone else's opinions and disbelief in facts they themselves do not know or didn't think up first are the bulwarks of such twisted minds. David Harum noted such differences among humans when he averred, "They's as much human nature in some folks as th' is in others, if not more." As they speed through life making everyone around feel uncomfortable, they don't realize speed is of no avail if you don't know where you're going. Their bluster hides their inadequacies because they believe what Lyndon Johnson said: “If you don’t blow your horn, somebody will steal it.”

The negativists, self-appointed experts on all things, are the classical misfits, malcontents and active miscreants who make travel with strangers hazardous to one’s health and well-being. Their bafflegab and sophistry belie the truth so necessary for friendliness. If you have a stone in your shoe or a burr under your saddle, cast it out. Though one would get more than a little satisfaction if such travellers would fall overboard or be eaten up by lions or sharks, such drastic behavior is hardly possible. The best solution is avoidance, which never fails to “boil their oils!” Seldom are such people sincere, which leads to Heri Peyre’s comment: “The primary condition for being sincere is the same as for being humble: not to boast of it and probably not even to be aware of it.” Insincerity is a major component of the sophistry which seeks to push down others to elevate oneself. It breeds distrust, and if a man asks, “Don’t you trust me?,” watch out. That man is not to be trusted in word or deed.

Lest one become too self-righteous about avoiding uninteresting people, with perhaps serious misjudgments about the aberrant personalities one inevitably must meet, it’s good to think of *Philipo* by Robert M. Shortell: “It takes an honest man to be able to laugh at himself and face up to his own folly.” So those of us who are unsure if certain people are really as bad as we may at first believe must at least try to understand them. Many tries later—all unsuccessful because the opponent baselessly thinks the same of you as you do of him—you recall Lew Alcindor’s dictum that “When all else fails, try a right upper cut.” However, remembering that the right words are better than a fight, and fighting is the domain of the unimaginative, simply going out of one’s way to avoid unfriendly personalities is the best solution to the problem.

I have noticed that many uncivil people are unsure of themselves, which leads me to Ralph Waldo Emerson’s thought that “A scared man is seldom rational.” Such people can become two-bit tyrants lashing out at others because of their own shortcomings. A personal bit of advice: When you take down people with whom you associate, you do not elevate yourself; you only make yourself look stupid. Some of these people are volubly talkative to a boring degree, so in listening you can double the quality by cutting the amount which you believe in half. Where such stupidity reigns, other fools will congregate and succeed. Hence, so often we see such mentally untidy people becoming friends with birds of the same feather.

If by chance one should try to correct statements by mentally aberrant associates or make efforts to protect men from their own follies, the earth would soon be peopled solely with fools, to paraphrase Herbert Spencer. People who live a third-rate existence generally don't care to be contradicted. So why give them the advantage of knowledge you have gained by hard work? Besides, advice unsought and free is worth NOTHING to an obdurate mind. Frank Gifford, that loquacious TV announcer of football games, once said, "Everybody's got to be somewhere." True, but why do we have to meet those somebodies who are difficult people when friendliness is so easy and so pleasant—especially on a ship where escape is impossible?

The *Morrissey*
Frobisher Bay, Baffin Island

Ignorance and Stupidity

A plaque in our motorhome has on it a phrase the origin of which I have never been able to ascertain. It goes like this:

SAY NOTHING
DO NOTHING
BE NOTHING!

It reminds me to consider the vast differences among human beings, especially concerning not only intelligence but also involving both mental and physical alertness. Attempts to ascertain the difference between ignorance and stupidity lead to a morass of contradictions. Try as one will, there are no simple answers as to why humans are so unequal. A six-fold view of the human condition may illustrate the difficulty:

They look but they do not see.
They listen but they do not hear.
They sense but they do not feel.
They talk but they say nothing.
They read but they don't understand.
They walk but they go nowhere.

Wherefore are they alive but not living or are they the living dead?

We tread on dangerous ground when we attempt to judge people. Intelligence quotient tests won't do the trick, although they may be a measure of how facile a mind really is. Chief Joseph was fond of observing that one should "Never judge a man until you have walked a mile in his moccasins." Many a stolid, unimaginative mind in a body with photographic memory and loquacious tongue may lead an individual to success. Hence we must realize that there are thousands of ways to be successful. Success cannot be a measure of the difference between ignorance and stupidity. Basically each person, in dealing with other persons—as all but the veriest hermit must do—needs to develop his own criteria for making judgments about people. After all, it's good to realize that half the people in the world are below average!

My own approach is to consider every ignorant person as teachable and most, if not all, stupid persons to be unteachable. Here, again, care is needed. There are many quite stupid people who work successfully with their hands, and there are many ignorant people who simply cannot learn to do menial tasks. Again a dilemma. Is that because some ignorant people do not choose to work hard with their hands, legs and backs? Many truly intelligent but ignorant people are real klutzes given a hammer, a saw and a ruler.

The reader who truly wants to learn all about stupidity must read Paul Tabori's marvelous book *The Natural Science of Stupidity*, published in 1959. The entire book is quotable wisdom. For the intelligent, opportunity comes many times; for the stupid, it seldom appears. It is not unusual for the stupid to make fun of things and activities that interest the intelligent. Genealogy is one of those fields of endeavor that leave mentally slow people cold. I heard one dolt tell my wife, an accomplished genealogist, "I don't want to know about my ancestors. I might find a horse thief or murderer back there." Another person, obviously intelligent, said, "If I find a horse thief or murderer among my ancestors, I'll know I have an interesting background." May such observations be a measure of intelligence or just of prurient interest in criminality?

Santayana, the intellectual, believed that "Skepticism is the chastity of the intellect." Even skepticism can be a detriment to discovery, which leads one into the morass of deciding between objective and subjective reasoning. To my mind, skepticism cannot be a measure of intelligence. One of the stupidest persons I know believes nothing told him is correct unless he thought of it first. Try as I may to avoid people like that, it is difficult because there are so many of them! Kahlil Gibran asked, "Where can I find a man governed by reason instead of habit and urges?"

So we must leave this problem as one so esoteric that it can't be solved. All people are interestingly and amazingly different. It can be honorable only to assume that all people with whom one associates are intelligent until they prove themselves to be stupid, whereupon avoidance is the best policy.

1966

Stubborn Versus Tenacious

Conscience may cramp any man's style, and the strain could break imagination. Uncertain idealism can gnarl the mind of an old-fashioned man. A stubborn man is seldom a man of good will, and a tenacious man may be doomed in a world of arrogant passions and ruthless compromise. Why are stubbornness and tenacity so different? The little *Oxford Dictionary* defines stubborn as "unyielding; not docile or amenable to control," whereas tenacious is defined as "clinging tightly; slow to relinquish anything; retentive." In dealing with students over a period of 30 years, I have found another vast difference. A stubborn person seems unable to change his mind despite proof of error—a sign of stupidity so flagrant as to preclude learning. I consider this a flaw which is not easily corrected in a person's character. But a tenacious person, though ignorant, can be taught and will then adapt to ever-changing conditions—a sign of intelligence. Tenacity may then be considered GOOD and stubbornness unhelpable or BAD.

Ralph Waldo Emerson, a self-actualizer who spent his life saying and writing things that seem eminently wise, put it this way, and I paraphrase: "Although we may change our opinions from hour to hour, basically we may be said to be on the side of the truth." Does that allow one to infer that a stubborn person is always wrong? Not at all, but one can harbor the thought that stubborn people are more than a bit stupid. There is much to be said for tenacity in efforts made by persons who know they are in the right. Just as the value of a pudding can be determined by eating it, a tenacious person who wins has proven his rightness many times in the face of great odds. But who can admire a stubborn man who is so often wrong? What parts do prejudice and preconceived notion play in stubborn rejection of the truth? I've also noticed another characteristic: stubborn people never apologize for their errors and seldom admit doing wrong. Tenacious people may hang on in fearful situations and win, but every tenacious individual I've ever known who was wrong could admit error, which, if it led to harm for another, resulted in apology.

One of the earliest things a young person should learn is to differentiate between tenacity and stubbornness in other people and in himself. Some people are never wrong, according to their own lights. Just as there is no person who will never die, so is it true that there is no person who is always right. The most dangerous people I know are stubborn people who have power over others. Their meanness halts progress, and their administrations are rife with errors of judgment. To join and agree with such a person doubles the ineptitude. Better to move along to different fields of endeavor than undergo the danger of losing one's own self-esteem to further a career whose activities result in frailties that cannot be corrected. A man who believes in nothing greater than himself is small indeed. If we are a part of all we have met, just so a bit of what we are is in all who have met us. Consorting with a stubborn man belittles the soul. There is an old saying in the oil industry that "He who hauls manure can't help but get some on himself." Roughnecks on drilling rigs substitute a four letter word for manure.

Demosthenes in 348 BC said that "We believe whatever we want to believe." From *Saints and Sinners* by William Williston one gleans this statement: "We often prefer established error to novel truth." Whence comes the prejudice that so typifies those of stubborn mien? To get to knowledge, one needs to garner wisdom from many people, observe actions that are successful, be willing to see all sides of an argument, and attempt to divine what makes people tick. Blind faith won't give us answers. One must come to the conclusion that the ignorance of tenacious people can be cured by learning, but nothing can cure stubborn stupidity. There is a well-known German proverb that says,

"Stubbornness is the energy of fools." Eric Severeid, that wise television commentator, believed, "To be ignorant in today's world is to be anonymous." The mayor in that wonderful book *The Milagro Beanfield War* typifies the stubborn man when he said, "If we don't know it already, we don't need to."

Finally, does tenacity lag and stubbornness increase with age and experience? Plato said, "For many the vision of the mind does not become keen until the vision of the body is past." Do we become less intelligent with age and therefore more stubborn? Must we become so set in our ways that we lose our tenacity and depend on stubbornly held ideas? Listen to your children and grandchildren, and, if you hear them well, you cannot endure the ignominy of stubbornness. Would that they could learn bulldog tenacity early and eschew stubbornness!

At work on Baffin Island
1943

Rule by Ruckus

There are some people who move across the stage, stumble on a loose board, get knocked out of step, and create a ruckus that stops the play. The pratfall has always been a thing of humor. We tend to laugh until it dawns that the prostrate form hasn't moved for a few moments. Rushing to help, the thespian sits upright. He turns a baleful eye toward your legs, pivots on his buttocks and kicks you in the groin. Pandemonium reigns, the rehearsal stops, and the director shouts "take five" while fellow players separate and wade into the slugging match. How can all this be? We're supposed to be logical, loving humans getting ready to put on a play, romantic, not Rambo-style. It's a matter of perception. The laughing angered the downed thespian whose boiling point is low because of his clumsiness. Back to order, the "five" is over, the villain who stumbled is properly villainous, the hero has a black eye, and the rehearsal goes on. But a little levity is usually a good thing.

One of the favorite tricks of a paranoid person who is in a position of power is to embroil everyone in an artificial ruckus to cover up his own mistakes. He feigns great anger over the slightest hurt, believes the whole world is against him, and woe betide the person who laughs as he recognizes such tricks of diversion. The answer is to avoid such persons. I once had an academic boss, a chairman, who told me he never brought important matters to the table at staff meetings. He would fill his unwritten agenda with items of no consequence, then make assignments of duties, and later change the rules with a ruckus of blame for mistakes trumped up about assignments correctly completed. This led to several of us taking complete notes at all staff meetings. Noting this activity, the chairman ceased holding staff meetings for long stretches of time until the dean became exercised that our department was embroiled in major controversy over the lack of leadership. This lack of logical decisions on major problems involving the department brought progress to a standstill, as the dean raised a ruckus that could be heard all over the campus. This gave rise to the concept that deans are truly the bastards in academic workplaces.

H.L. Mencken had this to say about ruckusing politicians: "The secret to success in politics is keyed to keep the populace always uneasy and in a constant state of alarm." Also, politics appears to be the sly craft of appeasing the voter without giving him what he wants. Rule by ruckus is the ideal bafflegab of dishonest politicians, and in no place is it better displayed than in the public utterance of ultra-liberal congressmen seeking to stop all developments under the guise of "saving it for future generations!" The answer should be conservation, yes, preservation of the status-quo, no!

Joussenel is credited with the saying that "The more one comes to know man, the more one admires dogs," but Jean Cocteau said much the same thing. Who said it first? I'm always intrigued by authors and book publishers who raise a great ruckus over supposed plagiarism when any one with common sense knows that whether you read it in a book, newspaper or magazine or thought a great thought *de nouveau*, no one has a right to claim copyright infringement on how words are put together. Years ago a friend of my father told him about teachers: "If you can, you do. If you can't, you teach." This was when my dad told his friend I was going to teach in a university. I think Dad's friend thought it up himself although it has been credited to Voltaire! I'm sure he never read it anywhere. Many of Voltaire's writings and quotes could probably be found somewhere in the works of Seneca, the Roman Senator who was a Spanish Jew. Moise Bendrao Levy Ayash, a philosopher who once worked as an office manager of an oil company in Portugal, queried me on Americans this way: "Why is it so many Americans study psychology, and yet they have so many mental problems?" I believe he was thinking of American consultants who get livid with anger if their theories are challenged. The milk of human kindness is thin indeed in people who go into high dudgeon at the slightest provocation. In earlier days such encounters led to duels—a ruckus that surely should never happen between thinking humans. Yet who has not met people whose very conversations are blaming and cantankerous, every statement being a challenge to anything that is said? Avoid them. Let them stew in their own juices. Don't let them stew yours!

William James noted that "The essence of genius is to know what to overlook." Capacity to deal with an obdurate person is a mark of wisdom. Whether wise or not, I once told a cantankerous boss, "I can get along with the sorriest son-of-a-bitch in the world if I can figure out what makes him such an ass." It went right over his head, for he knew not to whom I was referring.

My brother Owen
September 1939

Smart Alecks

The world is full of them—in all nations, among all nationalities, within universities, colleges and schools. In fact, in any group larger than ten in number, we will find one S.O.B. and another will be a smart aleck—not necessarily the same person. Their egos are high, though ego itself is not a bad thing, for ego is an important ingredient of success. It's their blatant demonstration of smartness, many times when intelligence is lacking. In *Time* magazine of December 25, 1978, John Skow stated, "Too much rightness shuts off debate and stifles the thought processes." Instead of building bridges between people or groups, one person can throw akilter any hope of progress in frank discussions if that person maintains that he is always right. And when two smart alecks are in the same group, everyone else might as well go home. In no other group is this so deadly as on a jury or on a committee. But a smart aleck deserves study, if only to find out how he ticks. Perhaps it's temporary, a bias brought on by a bad breakfast, a bad night's sleep, a fight with his wife, the loss of an argument, or a serious challenge to his judgement. The person who is always self-perceived as right in all situations must be shunned, especially if it is a permanent bias against the ideas of others. There simply is nothing so dull as a self-righteous person, particularly in discussions of religion or philosophy. But perhaps God preserves such error so we may remain interesting and human. The low-brow approach is "If there ain't no dumb people, what else could we bitch about?"

Disraeli once said, "It is easier to be critical than correct," and Harry Truman maintained that the only things worth learning are the things you learn after you know it all. But you can't learn by talking, only by listening. One of the characteristics of the true dyed-in-the-wool smart aleck is that he talks incessantly. In Old Lyme, Connecticut, an anonymous note in the local newspaper stated this: “Those of you who think you know everything are annoying to those of us who do.” So who is the smart aleck? People such as those above or those of us who think others are guilty? In any group preparing a plan of action, there seems usually to be an abominable “no” man. One can’t help but be wary of critics, for they are so often wrong. Another failing of the smart aleck is that he doesn’t believe in serendipity. “Heaven forbid,” says the smart aleck, “I am in control. I’m right. I’ll make my own happiness and discoveries through sheer intellect.” How dull! If you’re afraid of being lonely, don’t always try to be right!

How best to handle a smart aleck whom you cannot avoid, such as a boss, a grandson, a person who has something you want, or a brother? Running away is a drastic solution. Avoidance another. A call for him to shut-up rolls off his back like water off a duck, and, besides, it might be dangerous. Convincing him he is in error takes great diplomacy, the type where you lead the miscreant to the knowledge that he himself is wrong. Perhaps no man in history was more successful in that endeavor than Benjamin Franklin. Adroit with words, using the technique of possibilities, probabilities, even parables, he, more even than Thomas Jefferson, led the most famous band of individualistic revolutionaries to agreement on the words that started our nation to greatness.

A great deal of what leads to calling a person a smart aleck is misunderstanding, which led to Senator Sam Irwin’s famous quote in 1973: “I know you believe you understand what you think I said, but I am not sure you realize that what you heard is not what I meant.”

June 21, 1992
Calgary, Alberta
Canada

Achievement

Mrs. A.J. Stanley had this to say about achievement:

"He has achieved success who has lived well, laughed often and loved much; who has enjoyed the trust of pure women, the respect of intelligent men, and the love of little children; who has filled his niche and accomplished his task; who has left the world better than he found it, whether by an improved poppy, a perfect poem, or a rescued soul; who has never lacked appreciation of earth's beauty or failed to express it; whose life was an inspiration, whose memory, a benediction."

There can be no more philosophical treatment of the meaning of achievement. Every man, woman and child who seeks acceptance and approbation for actions has a self-appointed standard which, when achieved, shows success. Eugene Csernan, one of the original astronauts, said, "There is nothing impossible in this world when dedicated people are involved." In that statement lies the key word: dedication. Failure to achieve a goal leads me to Abra Iban's thought: "Nothing is inevitable until it happens." This applies if failure is allowed to ruin the will to achieve. As to approbation as a goal to achieve, we should listen to Marcus Aurelius Antoninus: "Always be sure whose approbation it is you wish to secure and what ruling principles they have." Horace, that civilized Roman among many, said, "The man who is just and firm of purpose can be shaken from his stern resolve neither by the rage of the people who urge him to crime nor by the countenance of the threatening tyrant." Justice Oliver Wendell Holmes put it into modern terms with

"If you want to hit a bird on the wing, you must have your will in focus...every achievement is a bird on the wing." In seeking to achieve a goal, it is good to remember Thomas Jefferson's advice to a young friend: "Whenever you do a thing, though it can never be known but to yourself, ask yourself how you would act were all the world watching you, and act accordingly."

Every person who has achieved a number of goals knows others who have not. From those people we learn what makes for failure, and, therefore, conversely what makes for success. Elsewhere I have noted that failure need not be permanent.

A former student of mine, one of Carlson's Marine Raiders in the Pacific during World War II, had an ambition I thought only tributary to his effort to become a professional geologist. Although I didn't discourage him, he asked my advice, and I told him that I could see little benefit to his career. He wanted to go to the headwaters of the Orinoco River in Venezuela and find gold, diamonds, rubies, sapphires, and other precious stones. A tenacious student with an ambition, he delayed academic work, went to South America, and came back with literally a bucket full of valuable gems in the rough. With renewed vigor after the achievement of a dangerous goal among hostile natives, snakes, mosquitoes, crocodiles, and God only knows what else, he settled down to academic achievement which led to his becoming a millionaire as an independent oil producer. What this proves is that a professor's advice is not always the best against the burning desire of a student to achieve a goal.

Achievement, though often geared to the eyes of the beholder, must in the final analysis be what the achiever himself feels he has accomplished. Other students of lower potential have many times recognized their shortcomings and turned their efforts to lower objectives to achieve notable successes. To those I give the very highest credit, for they have achieved happiness in their works and goals of value for their families and the people whom they serve. What better measure can there be of achievement?

On vacation in Mexico

A Life to Live

It is not unusual for an imaginative person to model himself after other people whom he admires. All of us are ever-changing as the result of what we observe among those we meet. It behooves us to meet the right people, and that takes judgment seldom found in the young. For every quotation that spells out a guide as to whom we should emulate, there are several that say, "Be yourself, or how would changes for the better ever occur?"

Without being too specific, allow me to assemble some comments about the difficult business of living a life of observation, whether satisfactory to others or not. Failure in any endeavor depends on who—yourself or someone else—observes it. The freedom to fail, as often stated, implies that there is an equal right to succeed. It is all a matter of right or wrong choices. As soon as a bit of maturation comes to a youth, he realizes the concepts of yin and yang, black and white, good and bad, but the real discovery is two-fold: first, all gradations are possible, and, second, at times every person is bad and at other times good, with a mix or gradation most of the time.

Depreciators of other people's choices and activities seldom see their own shortcomings; when these types look in a mirror, they see perfection. Obviously their standards for themselves are abysmally low. Little progress is made in one's own life by doting on the apparent failure of others. Look within! Many egomaniacs maintain that they have great ability to manage the lives of others, yet they make complete asses of themselves in their own lives. I've always found it difficult to boss people around, given the chance to do so in business, scientific societies and the United States Navy. Why? Because I have enough trouble just bossing myself around. The answer may lie in those who are martinets toward their peers and underlings. They cover up their own ineffectuality by running the lives of others. Best to avoid such people, if possible.

Understanding prejudice in others is another difficult problem. My mother thought all humans to be equal. My father didn't like Catholics, Jews, Negroes, women drivers, doctors, and "slow" people. Yet three of his best friends were Catholic businessmen, and three others were medical doctors. He put up with Jews in business, but he always maintained that no Jewish person could ever best him in a business deal. As for blacks, he was proud that not one black person lived in Holmes County. As for his concept that women drivers should be outlawed, he was wrong, of course. Observing his prejudices was the major reason I detest prejudice and am apt to raise heck with friends and acquaintances who make snide remarks about anybody.

It is when one considers courage that one sees such great variations in humans. In a foreword to John F. Kennedy's book, *Profiles in Courage* (written by a ghost writer), Allan Nevins says, "No man without character is consistently courageous." Character is often mistaken for reputation and vice versa. How does one determine character? How does one measure it? Everyone is at times bad and at other times good, in some things bad, in others good. One of my great-grandfathers was a brilliant Amish-Mennonite Bishop who worked himself to death at the age of 54. He overdid physical labor and was revered by the Amish. He was, according to family anecdotes, more than a bit addicted to the grape. Was he a courageous man of good character or a bad one? He and his intelligent, beautiful wife, a full head taller than he, raised five stalwart sons, all seemingly of average intelligence, mostly successful in that when they died they were not in debt. One of those sons was my courageous grandfather. Were these people of good character? At least not one of them ever spent time in jail!

One of the great difficulties in living a good life is finding time to be alone. Time by oneself is precious. Time to contemplate, time to plan without hindrance by the prattle that goes around one in the family or among friends and fellow workers. Yet conversation and personal interaction are necessary as a wellspring of human development. All good things come in moderation. I have one friend, an engineer, who simply can't stand to be alone. He is never lonely when he is in a crowd—as many people are who are loners. He has found his *modus operandi*: be with people all the time.

Some of my Portuguese friends tell me that any man who doesn't like bacalao doesn't like women. Their idea of the way to live is to eat cod and love women. Another is a scofflaw, and, during the reign of Salazar in Portugal, the secret police tabbed him as a revolutionary simply because he railed at dictators. A highly successful engineer, he now lives in Belgium, a democratic monarchy, happy in a life quite different from his difficult days in Angola. He found his niche, his way to live.

Norman Vincent Peale, in his article, "Trouble: Whetstone of Life," said that "Adversity is the abrasive that gives a sharp edge to courage." He advises one to face up to problems as a way of life, to take a hard look at oneself, to take some kind of action, not to be afraid to seek help, and not to fall in love with one's problems. If poetry is what Milton saw when he went blind, how can those who see not find a satisfactory way to live? It's a matter of scale, intensity, desire, opportunity, ambition, and having been born with the right genes.

Mark Twain, that sayer of many things who had some shocking observations on the frailties of humans, is oft-quoted so: "Always do right. This will greatly gratify some people and astonish the rest." The sophisticate—impatient with naiveté, distrustful of people while being himself untrustworthy—considers truth harmful if a lie will do better. But there is a way of life, for whatever reason, that many people follow: do good; always tell the truth; if you can't say anything good about a person, then say nothing; best to be thought stupid by being silent lest you open your mouth and be found out to be a fool; and, overall, leave the world a better place (because you've been here). Who cares what other people think so long as you live your life your way and hurt no one but those who are evil. However, don't waste any part of living looking for evil. It's all around you. Just be sure there's no evil in yourself. Trite to say, but there are many ways to live a life. Make sure yours fits who you are!

Disappointments

Harry Truman once said, "You can't always do what you'd like to do, and the sooner you learn that, the better off everybody is." Many small boys, on seeing their first fire engines, want to be firemen when they grow up, or airplane pilots or circus aerialists. Little girls think they would like to be movie stars or mommies with their own babies, which often happens too soon. Looking back in my own life to a boyhood on the farm, I had hoped some day to become a race car driver because my father often had taken us to car races on dirt tracks. Once he took us to a steeply banked wooden track near Akron.

My earliest wish was to fly airplanes. As I grew older and learned to fly, the desire was to be a transport pilot. Once one attained 200 hours of flight time, one could take the old Civil Aeronautic Authority examination and earn one's professional ticket, then he could get a job flying passengers on barnstorming tours which were so popular at county fairs in the late 1930s. But it was never to be. When I had the 200 hours, I was deeply immersed in becoming a geologist. Not getting the transport license was a great disappointment, but I consoled myself by thinking that I didn't really want to be an aerial bus driver and fly for an airline. After a thousand hours of flying, I still get a twinge: Why didn't I go for that transport ticket?

Academic achievement became a goal, but I didn't even hear about the Phi Beta Kappa academic fraternity until late in my junior year in college. My first half year in college was marked by straight C's because of my strong feelings about a New Dealer economics professor and lack of interest in several other courses. As a result my final average in college was a B+, and one needed an A- average to be considered for Phi Beta Kappa. This disappointment was short-lived as life became more and more exciting.

During the first year in college, all freshmen were required to pass a rigorous test in English grammar—a truly old-fashioned requirement which I flat flunked with a grade of 70, the passing grade being 75. This disappointment was great, for I had gotten good grades in grammar and literature in high school and had received high grades on essays and short stories. Later, I did pass that cursed examination and was allowed to go into my sophomore year. The disadvantage of not having spoken English until grade school may have been the cause of failure, but the good grades I received in German, after two years of French in college, ameliorated somewhat the shame of flunking the freshman English grammar examination.

Although I ranked in the top 10 percent in the International Wakefield Model Airplane Flight Contest at Wright Field in 1931, disappointment at not winning first place was my lot. It was only slight consolation that I was the only boy in our region of Ohio who built flying models. No one had taught me how to design and fly rubber-powered airplanes. When I won first place in the flying contests at the Mansfield Air Races that same year, the manager of the air races absconded with all of the entry fees and gate receipts, so I lost the $25 first prize—a disappointment not at all matched by the thrill of learning that that crook finally made it into the penitentiary down in Columbus.

One of my biggest disappointments was not making it into the major leagues as a baseball player nor winning my college letter in football or basketball.

Failures seemed to stalk efforts to be tops in something that seemed to be laudable and possible to attain. It never occurred to me to want to be rich, though I wanted the freedom to do as I pleased. I wanted mainly to travel and see the world. This became so much more a viable objective than being a prize-winning farmer or the best roofer-carpenter in the business in Ohio. When I found geology, combined with geographical exploration and especially the search for valuable resources in strange lands, I found my niche. Yet it seemed that so many were more capable as scientists. Not to worry. I'd do the best I could, and let the chips fall where they may. And they did!

In retrospect, I think it was my numerous failures and the resulting disappointments that burnished ambition toward achievements which I have made. When a man feels that he has been lucky, that is a kind of success in itself.

With one of my model airplanes
1933

Me (far left)
with fellow geologists
near Mt. Taylor, New Mexico
1982

The Positive Equation

What is it about people whom one immediately likes upon first meeting? As a person who has lived a very long time in milieu ranging among roofing crews, college classes, university courses, seismograph crews, exploration departments of oil companies, university faculties, flying clubs, scientific societies, groups of Naval officers and sailors, and travelling groups (voyages, outback safaris, group tours, etc.), I found that some people start out as likable characters with bright personalities. They listen attentively, do not interrupt conversations, have interesting things to say, smile a lot, laugh unabashedly, and radiate friendliness to an extent one finds unusual in most people on first meeting. Of the literally thousands of people an active, interested person meets in a life time, perhaps several hundred such paragons of friendliness are remembered with glee. Many become friends worthy of continual contact. No ulterior motives exist to do so, only a good feeling of friendship; they are the people who make us realize the dastardy of unlikable people of the negative equation.

The people who make us happy at first meeting are those who make us feel superior by paying rapt attention as we regale them with outrageous stories. Perhaps that's too strong an appraisal. If they but listen with care without looking past one or butting in, that is a kindness lending to the positivity of the human encounter, the happiness of the equation. People with laughing eyes, who smile rather than grouch along, have interesting thoughts fearlessly expounded and give the listener time now and again to agree or even to disagree agreeably. Benjamin Franklin, that connoisseur of diplomacy in personal relations, never flatly disagreed with listeners but would gently suggest that he thought differently and gave cogent reasons.

Life is happily filled with many more people we can like, or love, than with grouches. Of course there are many for whom one can harbor no feeling one way or another except to be civil and learn what one can; at least one need not actively avoid such neutral personalities. Most likable people have few prejudices or, at least, are intelligent enough not to make prejudiced statements on first meeting. My father's dictum was "Let there be no strife between thee and me, for we are brothers." He was a man who drew friendship out of even the meanest of men by the magic of listening, even though he may have wanted to carry the conversation himself. He made people feel important, without having an ulterior motive. But let a man cross him, and he could use vindiction as a sledge hammer through conversation so cutting that he was feared.

Is good humor that makes one like another person on first meeting some kind of emanation from a soul? A negativist will query "Is it a tragedy that there may be no soul after death?" A positivist will counter with "It is a supreme tragedy if a man's soul dies while he is alive!" It takes care and a benevolent perspicacity to draw out a person who does not exude friendship at first meeting. Is it shyness, stupidity, or has that person already decided, based on your looks, that you aren't worthy of time for conversation? Disturbing as such a decision might be, there is an old trite aphorism which says that "Still waters run deep." To get such a person to open up is a real challenge. I am reminded of what was said about Dr. Delbert Lean, speech professor at the College of Wooster: "He taught the timid to speak with confidence and the confident to speak with care." Dr. Lean had such a shining personality that he could draw blood out of any shy turnip.

One can measure the worthiness of a person by the size of that which irritates that person. Anyone you meet who flies off the handle at inconsequential things is usually unworthy of your friendship. There *is* value to equanimity, but if a person is too serene, perhaps there's a problem with the libido. A well-balanced personality usually has a sunny disposition and radiates confidence without being overbearing. Clearly, such people are worth knowing. Ernie Banks put it this way: “We have to generate our own happiness. If you’re not happy in one place, you probably won’t be happy anyplace.” Anyone can be a pleasing personality where the going is smooth and the breaks are coming along, but the test of a buoyant person shows up under stress. When you can say of a person you meet “He fills the room with laughter,” there is a person worth knowing!

Marshall Islands
1944 - 1945
I am on the left.

Minimum Lives

A person who has no deep down feelings, but only unbridled desires, is indeed an impoverished soul. To live a minimum life is the lot of many disadvantaged people who live only from day to day, gratified by the satisfaction of unimportant things, and have less important people as friends. Such witless ilk these mortals be. They do not live by their wits, for they have no wits about them. They lack thought for the morrow, satisfied with the feelings of the day. Happily, they make no progress other than eating, working as little as possible, and sleeping late in the morning. They deserve careful study. It's amazing how long these people live, they who live such minimum lives. They must know something which many of us, not as relaxed, do not know or comprehend.

Those who are slaves to their emotions deserve slavery. Freedom is the lot of those who are capable, who eschew minimum life. Albert Einstein maintained that "Our times are characterized by the perfection of means and the confusion of goals." Abraham Lincoln observed, "God must have loved average people because He made so many of them." I do not know what an average person is like. As a professor one soon learns that students just naturally fall on a bell curve, which may have little to do with intelligence and everything to do with ambition. Many people eschew intellectual arrogance on the pretense that wisdom displayed is elitist, to be avoided at all cost. If one is to learn by example, the characters who lead minimum lives are not good models.

But wait. Who can endure a person who is obviously always right or thinks he is, one who does not enjoy some common pleasures such as reading the funny papers, going to a burlesque show or drinking beer? Ascetic people seem to look down their long, thin, sniffing, aquiline noses at all but the obviously elitist. I know some wealthy playboys who lead the most minimum of lives. Harvard had a goodly share of them, and I suppose Yale and Podunk University do also. If talent is the present tense of past effort, then one must observe that many people who lead minimum lives have not had great models to follow from among their ancestors, their parents and their friends.

We call them scatter brains because so many minimum people talk too much. Stephen Leacock advised, "When holding a conversation, be sure to let go of it once in a while." Even hints and not listening will not stop such minimum people. Many of them seem to grin a lot. They smirk at anything that's said when there's nothing to smirk about. Derisiveness is their stock in trade when they can't or won't understand. To try to teach such people is a lost cause. Let them hang themselves with their twisted views of life. Henry La May once observed that "There is nothing more frightful than ignorance in action." Anyone who thinks education won't lift a person above a minimum view of life should see what ignorance creates.

Perhaps it is easier for many of us that so many misfits, miscreants and malcontents live among us, as long as we aren't required to bed with them. We owe them one debt of gratitude: they are examples of how we never want to be. How does one avoid those who are below average in most activities, or must one forever waste his patience with the severely clerical mind in low and high places? Honare Tracy once commented "The worst thing about a pig-headed man is there are so many of them." Shakespeare put it rather more elegantly: "Man, proud man, drest in a little brief authority, most ignorant of what he's most assured."

If it is true that the measure of a man is the size of what irritates him, then my appraisal of people who lead a minimum life may be thought to reflect on the possibility of my having lived a minimum life. I hasten to assure the reader that my analysis is not made through irritation but through a desire to appraise the causes of all human conditions which so greatly divide us. One characteristic of these minimum people is their alarming way of bragging, and that does irritate me. Yet it must be admitted, as Dizzy Dean said, "It ain't bragging if you can do it." Tom Landry put it this way: "You don't brag until you've done it." One can become wise if one observes what happens because of lack of wisdom in oneself and in others. From a braggart one learns little, except to stay quiet.

Gerhard Söhnge with my sister, Carol Ann
1939

Gerry and me as geologists
forty-seven years later in South Africa
1986

Failure

Failure in whatever light is generally in the eyes of the beholder, and there are tremendous grades of failure according to who you are. Graham Greene said, "Failure is a kind of death." Failure is relative and is usually caused by adversity, whatever the cause, whether self-generated or caused outside the ken of an individual. Sir Francis Bacon—of whom it was said that he was the last man to know all there was to know in his time—believed that "The good things which belong to prosperity are to be wished, but the good things that belong to adversity are to be admired." Many men and women go from failure to failure, trying again and again in the difficult areas of life. Henry Ford is an example. He said, "Failure is only the opportunity to begin again intelligently." He probably got that idea from his friend Thomas Edison.

It is common to hear the phrase among men of means, "I picked myself up by my own bootstraps." A vacuous thought, as every man or woman who became a success after many failures did so with help from others. Jimmy Stewart was emphatic with "No man is born to be a failure. No man is poor who has friends." Therein lies the seed of success after failure. Every failure who has become a success among humans had help from someone or from a condition created by someone, perhaps unknown. Peter Benchley, in *The Exploration of the Sea of Cortez*, knew from experience what he was talking about: "Failure is just as important as success because you had to fail in order to know that failure wasn't worth fearing."

Every person alive has failed more than once and surely must realize that failure may be an orphan among hundreds of successes. Elbert Hubbard believed, as does any successful human, "There is no failure except in no longer trying." Failure does not create enemies if it hurts no one but the failed, yet it causes the loss of erstwhile friends. One of the saddest situations among humans is that of a man of sheer bad luck being left alone as so-called friends abandon him like rats escaping a sinking ship. I have five friends who were notable failures—three now gone—but those of us who knew them stood by to help, and of the two who are left, one has become a notable success, whether by our help we do not know. We credit his own resourcefulness, as his friends stand by. I know of others who failed, achieved successes with help, and became obnoxious with pride of achievement. Such are among us, and they seldom realize how we now shun them. Some even became enemies, not realizing that their failures were of their own doing. They blamed everyone else but themselves. With such ingrates it is difficult to follow Abraham Lincoln's observation: "Am I not destroying my enemies, when I make friends of them?" The effort to be friendly to the ungrateful seems to be a lost cause.

We are not created equal but different, and to ascertain what is different about failure is most difficult. Certainly smartness may be a factor, as is adaptability. To pick yourself out of the mud and fight on is admirable, but a goal must be in sight or be had in mind. It was said of Daniel Boone—a notable failure, driven into the forests by his inability to get along in the civilized society of his day—that "having lost sight of his objective, he redoubled his efforts." But failures caused by what are called "Acts of God, flood, fire, war, and famine" (not one an act of God!) are looked upon as unfortunate. The failures created by ourselves get short shrift by the successful, perhaps unfairly. It is always a temptation to blame failure on stupidity, which can't be helped, rather than ignorance, which can be erased through observations, reading, hard work, and goals not yet acquired but firmly desired. I know a multimillionaire oil man who drilled 36 dry holes in succession before he found one of the largest oil fields in Montana. His goal was clear: to find an oil field. Don Meredith believed, "It's amazing when you're good how lucky you get."

Successful men and women are fond of saying that luck follows anyone with the greatest knowledge. Not necessarily so. I know some very smart people—OK, intelligent people— so dominated by aberrant emotions and deviant behavior that they have never been anything but failures. In one case a young man inherited a vast amount of money and could afford failure. He went to several universities, tried several careers, never finished anything, and died an embittered old man in his 60's.

So what does one make of failure? The basis to try again. Never make sport of a failed man or woman. They are too numerous. Some are too proud to accept help, others too stupid to succeed. It does not seem fair that for some to succeed many must fail. Anyone can fill some satisfying niche if he but try. The ultimate solution is death, and it comes to us all.

Chicago World's Fair
1933

Great Thrills

A life well-lived has in it many thrills, not necessarily of great achievement but of interesting activities which may seem humdrum and mundane to others. Who can forget such deep down thrills: learning to ride a bicycle, the first ride in an airplane, the first time a model airplane designed and made by one's own hand flew a sustained flight, the first kiss of a lovely girl, that first basket in a high school basketball game, the hit that won a highly contested baseball game with a rival college, your first home run in college baseball, graduations from high school and college, landing the first job as a scientist, the first pay check, the honeymoon, birth of a first child, winning the first election in a national scientific society, completion of the first oil well in which one has an interest, the first standing ovation by one's peers after a speech, first solo flight in an airplane, the first time driving a car alone, the first view of the Grand Canyon, winning a commission in the military, the first ride down a steep hill on a new sled, catching the first fish, the first time a rabbit was bagged with a new shot gun, the first flight of a homemade kite or boomerang, your baptism in a church, your first sexual encounter, and so on and so on!

Obviously such a litany is tedious for the sophisticate, or for the individual who is so blasé that nothing is exciting, or for a person who has experienced no memorable thrills. I have often asked friends this question: "What is the most thrilling thing that ever happened to you?" The answer is always surprising but believable. The next question is the one that stumps everyone: "How exactly did your body feel when the thrill occurred?" Most people simply do not remember! The stomach to genital area is the seat of all emotions, not the heart as poets would have one believe. There is an utter feeling of joy, an exhilarating tingling that begins there and suffuses the body. It is akin to another devastating feeling and burning sensation in the anal area in times when one is in great danger, such as a forced landing in an airplane, a near miss in combat, a searing automobile accident, or the death of a relative or friend.

Are there dolts alive who have not had such feelings? If so, are they alive? Beneath all the thrills involving special occasions or special experiences is the vast everyday joy of living. Old people have told me of their simple joy at awakening each morning. Is that a measure of when one is old? Or do older people get so beaten down by life that nothing thrills them anymore?

A measure of our sensitivity to great thrills is to consider our great disappointments. Many a very happy feeling is as positive as a deep disappointment is negative. Equanimity is a laudable characteristic, but one must be careful not to mistake serenity for senility or vice versa. There is a serendipity involved in many of the thrills in a life of many experiences. The word serendipity itself is a surprise, for it is a word coined by Horace Walpole from the Moslem name *sarandub*, the ancient name of Sri Lanka. Walpole said it meant "Aptitude of making happy and unexpected discoveries." To be surprised by unexpected thrills is the stuff of an exciting life. Some of what thrills us is due to our corruptions and folly, which led Erasmus to write a tract titled "In Praise of Folly" and Robert Frost to write "I like a little corruption myself, especially if it's amusing." Robert M. Shortall in *Philipo* said,"It takes an honest man to be able to laugh at himself and face up to his own folly," which makes one realize that a thrill derived from folly is truly accidental yet can be a vital lesson. Jack Abernathy thought that "The best things in life are free, and unexpected." Serendipity!

One must come to the defense of crazy moves in one's life, for they often result in wonderful happenings never to be forgotten and some spectacular failures. From thrill to abject disappointment is about the same distance as love to hate. Zorba put it this way to a conservative friend: "There's one thing you lack: the madness to be free!" With freedom to do unusual things and go to unusual places, the doors are opened to great thrills.

The world is full of spoilsports and people who say, "Yes, but..." Mention of a thrilling event to such a person is like talking to an oracle who replies, "But you should hear what happened to me!" To those misanthropes no grand experience by anyone else can touch their own. Mention your wonder at a chance meeting with a movie star, and you will be informed that he has met the President of the United States. The only solution is to avoid the braggart entirely and always. He who enjoys not the enthusiasms of others is unworthy of friendship or conversation. One must be watchful not to become such a spoilsport. We learn how to tell of wonderful experiences by listening to people who do not exaggerate when they tell of their own great thrills. To wonder is to live a full life filled with surprises.

Lt. Commander Sherman A. Wengerd
USNR

The Equal and the Unequal

It is in the area of human creativity that one finds the vast differences among individuals. People may by law be considered to be equal, but equality is a vacuous concept. The gene pool among all humankind is so widely disbursed that the wag is liable to say "There really aren't enough good genes to go around." One thing is certain: inequality in creativity, whether involving ideas or things, engenders jealousy in those less capable than the creators. In the sense that all of us have the same number of hours in the day, days in the week, weeks in the month, and months in the year, it behooves us to consider equality and inequality in the light of contemplation and work finely balanced. Some people are great in creating with their hands, a throwback to the time simians first stood up and found that the upper limbs had hands with fingers, no longer necessary for locomotion. Others, by projection of past experiences, create with their minds, learning from failure and superior extension beyond the normal thought necessary to stay alive. Freedom from routine is an ultimate liberty that leads those most capable of thought to go beyond equality with their neighbors. To put it as one pundit did, "Some are more equal than others." Al Unser, a four-time winner of the Indy 500, put it this way: "You can't be a winner if you want to be a quitter."

Egalitarianism has led to a new type of American, convinced that all should be equal in opportunity, but some of these Americans maintain all should have equality in things, whether earned or not. They say, "Life is short and difficult. Why should some have more than others?" This leads straight to the Communist credo, "From those who have to those who have not." Forgotten by most socialists is the idea that those who work the hardest and the longest, utilizing their entire mental capacities, should have their earned share of the spoils. To be required by law to share what you have earned with those who have chosen not to work and think is a sure way to create great losses in the American initiative. Put another way, there is something wrong with a system in which efficiency is punished by confiscatory taxation in order to care for the inefficient. However, the Christian ethic is correct. We must take care of the lame, the halt and the blind, those unable to care for themselves. The able-bodied, who will not move to a new location for whatever reason to find new jobs, should not be cared for as though they were incapable. So it seems that much inequality is self-inflicted by wrong-headed emotion or illogical desires.

When we are born, we are all equal, helpless and unclothed. When we die we are all again equal: dead. But in-between, here in America, we are equal in only one way: each of us, after a certain age, has one vote. This life interval is our one great known time of opportunity. How we handle it, prepare for it, the decisions we make in it, and, to an extent, how we view luck as being prepared for anything determines success. My father, more than once, told me I could do anything I was smart enough to do if I worked hard enough and didn't allow little-minded people to get in my way. I soon learned that inequality was to be recognized, provided you sought to be superior to those dominated by desires beyond their means. One of my mom's churlish friends put it this way: "It's heck to have a whisky desire with only the change to buy a beer."

If negative inequality is foisted on ambitious, hard-working people—as was the case with slaves in America—it is a wrong that must be righted. To deny opportunity for a capable person to obtain more than just necessities is wrong. John McKay maintains that "You deserve the right to be treated fairly, but you earn the right to be treated equally." There are more than enough people who are congenitally unable to be superior, but the opportunity must be there for them to try. Back in 1978, Robyn Davidson, the camel lady of Australia, said, "To be truly free, one needs constant and unrelenting vigilance over one's weaknesses." It seems then that a lot of inequality of the negative kind is self- inflicted: a marriage made too young, dropping out of school, general lassitude, ill health because of excesses, aims that are too low, satisfaction with a low status quo, undue desires, and a host of other reasons.

The early recognition of one's mental limitations allows choices to be made that lead to many avenues of success unknown to highly intelligent people, especially those who learn easily. I learned very early that there are hosts of people much smarter than myself, more capable mentally and physically, born to higher station with greater benefits gained without much effort. Does that lead to jealousy and seriously debilitating emotion? No! It hastens the search for that niche that awaits anyone who can do without so that more can be had later, whether it be education, riches, fame, or any good things of life. The Spanish have a proverb: *Con patiencia llegaremos a los cielos*, or with patience we shall reach heaven. The single most critical cause of inequality is the notion that "I've got to have it now." I know a family that has never been out of debt. They told me, "We've always lived beyond our income." The answer to this may be found in Salvador de Madariaga's view that "Inequality is the inevitable consequence of liberty." Nadi Reeder Campion's mother told her, "There are two ways to be rich: one is to have a lot of money; the other is to have few needs."

Honoré de Balzac held the view that "Rich people will never be capable of understanding the less fortunate." Perhaps it is too much to expect serious contemplation of the meaning of a mediocre life. Lives can be restructured, painful as such changes may be. The greatest possible opportunity for success comes to a person who is alert to the foibles of the unsuccessful. To avoid submergence in the cesspool of failure takes insight through education. *The Chicago Tribune* once published a quote germane to analysis of a major human condition: "It's a good thing most people are in a rut, for if they got out, they'd get lost." Ralph Aling, self-educated philosopher, put it this way: "If you keep your nose to the grindstone, that's where you'll always be, at a grindstone. Look up to see if you can't get someone to turn the grindstone for you."

Something to Look Forward To

Of the many things that inspire humans to create a better world, one is comfort. Another is to arrange one's life so that each day one can look forward to something. To some it may be a good book after the day's work is done; to others a long walk at dusk or the pre-prandial cocktail or two. For all who have reasonably good health and a willing partner, sex is the most looked-forward-to activity which drives wholesome ambition. To the criminal it may be knocking off a bank or murdering an enemy. Everyone must look forward to something, or else life has little meaning. Show me a person who looks forward to nothing, and I'll show you a nothing!

It has always intrigued me that many people I know cherish the challenge of helping others and how often that help goes unthanked. There's got to be a place in heaven—to use a wholly theoretical concept—for preachers, teachers and any who like the challenge of helping. Do we all work for rewards? Are rewards the sought-after results of our doing good? If so, the emotion is a hollow one, yet that plus a tax deduction may be what drives much of charity. To what may we attribute the feelings of a miser who gives money anonymously to poor people? If one wakes up in the morning, does one consciously think of the good things one can do that day, and if so, is that reward enough? Wanting is a measure of personality.

The best balanced persons I know love people, a reward in itself. Yet if one depends on the good will of acquaintances rather than oneself, disappointments become rife in one's life. To always turn the other cheek does not make for strong character. To recognize evil takes watchfulness, although trust is the coin of social intercourse. President Jack Kennedy once said, "Forgive your enemies, but remember their names." Enmity is what drives some humans to look forward to doing harm. Kirtley F. Mather, the geologist who had such a strong social sense, once said, "Amity widely spreading will drive enmity off this planet." Is it true that if we search for the good believed to reside in even the worst rogues and then nurture it, the bad will disappear? Hardly! To love one's neighbor requires that one love oneself, and complete trust is the medium of friendship. Therefore, looking forward each day to something makes one's life meaningful, but be watchful so that hollow emotions do not dominate our days. The human who looks forward to tragedy will reap tragedy, to joy will reap joy. Some can generate their own fate; others simply ride along to mental oblivion. Each one must find his own answers.

Bureaucracy and Progress

Bureaucracy is a necessary part of a democracy or a republic, or else who would do the slogging and tedious clerical work of keeping track of highly individualistic people. The ever-present danger is that bureaucracy becomes so entrenched that it takes on a life of its own—a hydra-headed amoeba that seeps everywhere into our daily life and poisons the will to be different or disallows progress in our complex lives. With a strong common language to bind us together, progress through cooperation has been made possible. Today, however, dangers are appearing that are more potent than they should be. There is coming a predominance of special interest groups whose initiations are based on socialism, religion, race, and different languages. That some will not be absorbed in the Judeo-Christian penchant for amalgamation into a cohesive American race is highly disturbing. The American Indians, first immigrants to America, are insisting on recognition for special treatment. This was not foreseen when, later, English and German immigrants established our democracy as a republic and then made the heinous mistake of separating main tribes into reservations. The Iroquois and many others have melded into the American stream of life, but the Pueblo Anasazi and the Navajo, plus others, perhaps rightfully insist on separateness.

The Irish potato famine near the middle of the 19th century resulted in waves of Irish migrants who have in part remained clannish mainly because of a Catholic background. Even earlier the Spanish migrations from pre-Mexico led to a Mestizo culture, predominantly Catholic, who wish to remain separate as Spanish-Americans. The name Chicano is an anti-American abomination. In another century they could become a dominant-force population. Already their clannish demeanor is having grave effects as they infiltrate the state bureaucracies of Texas, New Mexico, Arizona and southern California and seek special status. The descendants of former slaves lately have insisted on the terms "Blacks" and "Afro-Americans," and are beginning to have an effect on our bureaucracy by also insisting on special treatment in colleges and universities. They, thereby, inundate the lower to middle levels of American bureaucracy. Most recently, Cubans and Vietnamese who immigrated to America to escape communism have done a remarkable job of merging into our capitalistic system and will in time become more American than some who've been here for hundreds of years. Consider the Klu Klux Klan who insist that America be limited to White Aryan Protestant people. There is still much feeling against Catholics and Jews despite the basic tenet of religious freedom and separation of church and state; yet these two groups were among the first Americans in our early democracy.

Bureaucracy as a mode of making a living seems to be strong among the Irish and Spanish speaking segments of America. Yet all other groups—Jews, Catholics, Blacks, Indo-Europeans, and others—are seekers of places at the public trough. Bureaucracy is present in big business, big science, big education, and certainly best developed in big government, all state governments, the medical professions, and, more recently, scientific societies! The bureaucracy in Washington, D.C., may be likened to a flock of pigeons in a tree. Bureaucratic "technology is a way of multiplying the need for the unnecessary," as David J. Boorstin considered all technology. Bureaucracy tends to meddle, which led Bert Lance to say, "If it ain't broke, don't fix it." Gov. Edmund Brown said, "Bureaucracy is the epoxy that greases the wheels of government." Admiral Chester Nimitz gave Admiral Arliegh Burke a desk plate when Burke became Chief of Naval operation. It read "*Non legitimus corrondendum,*" which warned him of the civilian bureaucracy, or "Don't let the bastards wear you down."

Bureaucracy, whether in the Internal Revenue Service or any other branches of the government, is filled with unelected individuals who truly run the government of the United States. Most are difficult to replace when inefficient and cast a bad light on the many government workers who are efficient. There seem to be no solutions to the evils of entrenched bureaucracy, so we make the best of it. Sadly, many of our best citizens avoid government service, yet, strangely, many seek election to Congress. Makes one wonder at our priorities in a democracy.

Campers on the way to Mexico
1958

Why is it, and How Come?

In a long life, there are veritable hosts of riddles which don't seem to have answers, just as there are answers for which no questions are evident. The late Robert Roark, author of interesting books such as *Poor No More* and *Uhuru*, also wrote some of the most entertaining columns published in daily newspapers about the "why is it" and the "how come" of our lives, the little irksome things which happen so often and provide so many minor problems not easily solved.

Some people—in fact, many—are simply accident-prone, subject to the vagaries of life, and woolly-headed in the face of decisions easily made by most of us. There are people who just know that fate controls their every move and are fatalistic in the face of mistakes that grind their souls to unreconstructable bits. But fate can't really be blamed for the thousands of little things that go wrong or that can't be easily understood.

Have you ever wondered why the top half of an orange— that half we'd call the Northern Hemisphere if the stem end were the North Pole—is so much more difficult to peel than the bottom half? When you're at sea and there is flotsam floating by, why isn't there always some jetsam as well? How come some skinny people can eat like horses and not gain weight while those of us who only breath air and drink water get fat? Why do most tires go flat while the car is sitting placidly in the garage minding its own business? Then some joker comes along and says, "It's only flat on the bottom." The thought of murder comes to mind at such humor. Everyone knows puns are the lowest form of humor, yet why does everyone either laugh or threaten mayhem on the punster's head?

Why is it that a car or motorhome works perfectly

well all day, starts right off after gassing up, but then you park it at home at the end of the trip and the next time you want to go some place it won't start? The scariest setup is a car that dies going downhill. I've never yet correctly analyzed such doings. If I think it's a glitch in the coil or whatnot, it's a problem with the carburetor or a relay is shot. Thank God for mechanics!

Robert Roark noticed this one but had no answer: Why, as one gets older, does one have to get nearer to the mirror to shave in the morning—or in the evening for that matter? Why are so many heels and stinkers so successful, and why do many jovial, interesting people fail so often? Why do many good people die young and so many old farts hang around to become problems for us all? How come so many quotes by famous people are diametrically opposed to quotes by other people? Whom can one believe? And why do so many sayings by wise people get so mangled in their retelling? When a truthful tale is not believed and you hear it in its nth derivation, why does it mean exactly the opposite? Why do fillings cost more than dental extractions, and why do dentists suffer cappitus? Why must a tooth with a small hole be capped at $400 rather than filled at $50? Some kind of make-work must be involved. Why do baby teeth fall out? I admit it's a good idea for jaws to grow, and one would look funny with baby's teeth in a big, fat face. Why are some ignorant people so darn obtuse when it's so easy to be right about almost anything? Furthermore, why do some people have opinions and others have no opinions? Problems like that are shattering.

A shrug of the shoulder as an answer to a question is a rank insult. Why do people do it? How can so many people look so honest while they're telling bald-faced lies? Why do most children think a person over 30 can't be trusted? Is it because we don't listen to them in rapt attention? It behooves an adult to get down to eye level when talking to a child. If the bones and muscles are not willing because of age, have the youngster climb on a chair when you converse. You'd be amazed how nice the face of a curious child is when compared to the top of a head!

Why does a pen *always* run dry in the middle of a sentence? How come so many young people think older people don't enjoy sex? They'd be amazed at what we know and do! Why do young people think they'll live forever and old people talk about dying tomorrow? My father-in-law didn't think he was old at 88, so he married at 89, looking forward to living until he was at least 95. My own father talked about being in the late afternoon of life when he was 60 years old. Why such a difference?

Deans don't have to be the rats in the academic woodpile. Do they get that way because they must deal with selfish faculty members and recalcitrant students? There's an old story about a chap who applied for a teaching job as an assistant professor at a university. The president interviewing him said, "No, I'm sorry." The young man then asked, "How about as an associate professor or even a departmental chairperson?" No was again the answer. So the disappointed applicant exclaimed, "Well, I'll be a son-of-a-bitch," and the president piped up. "You're hired! We have a deanship open." Why are postal clerks so gruff and hostile? That's an easy answer. It may be because of surly postal customers. Of course, customers may have some right to be that way based upon what Ralph Schoenstein observed: "A letter recklessly entrusted to the United States Post Office is like a bottle on an asphalt ocean." Have you ever wondered why barbers always forget where you part your hair after it's cut, or why they want to trim your eyebrows?

Why does it always rain at picnics? Furthermore, how come ants can find a picnic so fast? There are many other frustrations at picnics. Witness this scene: Johnny and Mary, two neighbor kids at a picnic, both had to go pee. Mary pulled down her panties, and a mosquito bit her fanny. Crying, she moved and tried again, but the tall grass tickled her bare bottom. During all this Johnny was going up against a tree. Mary observed Johnny's ease and said, "You know, Johnny, that's a handy thing to have along on a picnic!"

What prompts people to say, "I'm descended from people who came over on the Mayflower"? On close questioning they don't know when their forebears came and don't know the names of their ancestors. Is that an effort to appear more grand than the facts allow? Why all this pride of ancestry when it's where you're going, not where you came from, that counts!

How come so many people, as they get older, hark back to the "good ol' days" instead of looking forward to a glorious age? When Art Linkletter wrote the book *Old Age is Not for Sissies*, he tried to point out the good things about growing older, provided you keep healthy and look at the aging process as a challenge. Of course a person who's 50 to 60 and keeps saying he's old is just that, a lost cause, and he might as well leave. One lucky thing: we tend to forget the bad and remember the good, but to wish to go back is a vacuous dream; it can't be done. Make the very best of today. If you must whimper, do it alone; the world is full of enough spoilsports. Why add to the misery already all about us? Why are there so many pessimists, sadists, masochists, and other asses who are dissatisfied with the lot they themselves created? Some people are so cantankerous and contrary that their only happiness lies in making other people miserable. The answer is to avoid them!

Why does money disappear so fast? How come we spend it willy-nilly with no thought for the morrow? I've never been able to understand people going head over heels into debt, paying interest to buy unnecessary luxuries to get a tax write-off. Many tax law revisions cooled that ardor, until some wise-acre and the banks advised people to mortgage their homes. If you want to have money to spend, make more or save longer instead of paying interest to banks and loan sharks. Of course, banks welcome giving loans to well-heeled customers who don't really need loans.

Why do French people talk at the same time during a two-way conversation? In fact, a Francophobic friend of mine wonders why the French still think they're God's gift to civilization. It would be wise of them not to criticize other nationalities for the faults they so plainly have themselves. It's a beautiful language, and they think they've given the world some of the earliest lively sexual practices. It really was the ancient Hebrews who thought of those techniques centuries earlier! Why be proud of things and ideas taken from someone else? In fact, why waste time being proud about anything?

These are but a few of the questions that still bother me—but not much—while I live an active life observing that most interesting phenomenon in the world: Homo sapiens, or us saps!

The Wengerd family after Dad's death—
Wilmer, Mom, Carol Ann, Owen, and Sherman
April 1966

Regularity

With a sort of detached amusement, a well-adjusted person watches the frenetic activities of disorganized fellow humans. The major benefit of scheduling one's time to create a pleasing regularity of things that must be done is that it yields abundant time to do the unusual, to think differently, and to try new activities that may have danger of failure, despite the potential of success. But such regularity requires a discipline which few strive to develop or do not have the tenacity to follow. Good habits, deeply ingrained, allow abundant extra time to dedicate efforts toward the unusual.

People who try to get other people to do things often make the mistake of asking an apparently busy man to assume new duties inimical to personal progress. Said another way, never ask a busy man to do something just to keep him occupied. The driven people of this world are the ones unable to organize their time to give themselves leeway for thought. Certain types of regularity must then be a blessing despite Ralph Waldo Emerson's dictum, "A foolish consistency is the hobgoblin of small minds." Oscar Wilde put it another way: "Consistency is the last refuge of the unimaginative." I don't think he meant all consistency. To be utterly consistent may well be deadening in all activities, but consistency in the humdrum activities of life allows for time to make golden new discoveries. In no other way can creativity thrive.

The Yoruba tribe in the Congo has two interesting tribal proverbs: "One should not go to bed with the roof on fire," and "To live with a humble man refreshes the spirit." Having one's roof on fire is highly irregular, but regularly keeping the buckets filled with water from the river, difficult a chore as that may be every day, allows one to put out the fire on the roof. The black people of the Congo have learned the value of such regularity. Further, they are humble before the fact that they must be ready to subdue roof fires started by invading tribesmen attempting to steal their wives, daughters and cattle. A confused person is of no value at such a time of danger. Once while exploring in Angola, I saw a Bantu man carrying a small dog around his neck, which seemed very strange. A Portuguese geologist explained that they regularly raise dogs for this purpose and always carry dogs around their necks while ranging through the bush. Why? A hungry lion will snap up the dog, allowing the Bantu to escape. Consistency in carrying a dog has saved a number of lives.

There are times when regularly scheduled actions are dangerous. These involve exactly timed movements in a city or taking exactly the same routes when walking, jogging or riding a bicycle. Our society is beset by bad people who break into homes, men who plan rape as they watch women's activities, and those who seek to mug people or to snatch the purses of defenseless women. To thwart such nefarious activity, one must think like a person intent on doing harm to others. Change routes, avoid dark streets, go with others rather than alone, and be totally irregular in your movements. During your times away from home, turn off the telephones, carry a whistle, be watchful in shopping malls and parking areas, and avoid carrying a gun unless you're fast on the trigger. In the state of New Mexico, if you shoot a mugger or robber, he'd better be carrying a weapon, and, if he's outside, drag the body into your house before you call the police.

A friend of mine in Flagstaff, Arizona, has a desk plate which reads, "The chaos about thee is but the confusion within thee." Ideally, a wealthy hedonist who cares not for time nor for doing any kind of good in the world can depend entirely on serendipity to fill his days with chaotically happy surprises. Regularity for such a person, who has little character, is not necessary. He has many friends who help him waste his talents and his money. When totally wasted he stands alone, a hollow shell. This happens so often as to be a simplistic truism. Many people regularly enjoy the company of a wastrel, but abandon him if he happens into dire straits. There must be some kind of lesson in such human behavior.

To each person there must come a balance between consistency and complete irregularity. Each must choose his own mode of living; what makes for success in one person may result in failure for another. To recognize the right course of action is a measure of a human's adaptability to changing conditions.

As President of AAPG
1972

Work and Achievement

It has always intrigued me that there are so many ways Americans earn their livings. Work provides the wherewithal to live, but it also uses up, for most people, a third of their lives during which they do not usually get into self-made mischief! There seem to be six kinds of work situations:

1) Those who work for others at regular jobs, eight to five, five days per week or some variant thereof;

2) those who feed at the public trough of government, whether in the postal service, education, politics, civil service, or whatever, at every level from village to Washington, D.C.;

3) those who work for themselves as entrepreneurs;

4) those who work as consultants at all levels in all fields;

5) investors, writers, artists, and inventors;

6) and dreamers who may work through the whole fabric listed above.

Of course, there are those who do not work at all, can't find work or are retired.

It has always seemed to me that people working for employers who pay out of their own pockets for jobs done work much harder and more efficiently than those who work for bosses who do not pay directly for work done. These include civil servants, teachers, congressional staffs, and those who work for corporations owned by the public. Another dichotomy highly evident is the division of those who produce objects and those who produce nothing tangible. An example is a wood worker versus a teacher. Simply put, there are those who work with their hands, those who work with their minds and voices, or those who work with their hands and minds. There should be a niche for anyone who wants to work! When you do get a job, don't quit looking for work.

The most rounded person at some time in his life, between early youth and retirement, has worked in all of the various ways in which one can make a living. Given the opportunities, or by searching out opportunities, working and adapting to the full range of experiences—particularly from wages for a job to salaries in a position—one can succeed in any field of endeavor. The Catch 22 situation which accrues so often—and unfairly—is that a job is available if you have experience, but how do you get experience if a job is not available? Does everyone have a right to work? If so, where do the opportunities come from?

Why is it that so many people hate how they make their living? Peter Benchley had it right when he said, "If something is not to your liking, change it. If it can't be changed, accept it." To put it more bluntly, if you can't or won't change either your work or the place where you work, then don't complain; the onus is on you. Complaining is a waste of time. The litany of excuses a person gives for failure to adapt to new situations of work is broad: "My wife doesn't want to move;" "We have many friends here;" "We can't find another place to live;" "I only know one kind of thing to work at;" "We like the climate here;" "We don't like to make new friends;" "We can't move all our belongings;" "We can't take our kids out of school and away from their friends;" "I haven't saved any money;" on and on and on. So it's easier to complain and bore everyone to tears with complaints than it is to do something about not liking the job. At many times in an active life, it comes time to fish or cut bait, take the bull by the horns, take a chance, plunge into the unknown. Ben Franklin pointed out one eternal verity involving wives: "If you would acquire property, first acquire a good wife." The same can be said for one's work: With a good wife, an adaptable helpmate, anything is possible. But a man without guts is doomed to failure, no matter how good his wife.

Richard Nixon, though his failing was an inability to face the music and say "I'm sorry," did make a memorable statement to the Congress in 1971: "Hard work is what made America great." Alberto Vadia, a Miami Cuban, must have been thinking about success at work when he said, "It is more important to be tenacious than smart. You can always hire somebody smart." Because freedom to choose one's work is basic in the American ethic, it is well to remember that such freedom is costly in time, money and effort. If tenacity is the prime ingredient in one's adaptation to any line of work, it behooves a person who works to equate time and money in relation to the amount of effort required to produce a certain result. To slog along, unaware of the value of one's time, is the fate of below average people. Charles Darwin, in July 1836, when discussing "The Value of Life," stated, "A man who dares to waste one hour of time has not discovered the value of life." Thomas Edison's philosophy was to start where other men left off. Joe De Yong, in 1927, said, "No man can make for me the clock that will again strike the hours that are passed." Such men knew the value of effort expended through time.

There is another side to this appraisal of work and achievement. There must be a balance among three major things in the life of a financially successful person. These are creativity, contemplation and relaxation. Stress only relaxation and one amounts to nothing; stress only contemplation and nothing gets done; with only creativity, one becomes a workaholic which leads to an early death. Every one of the truly wealthy men I have known divined the correct mix of those three activities and lived long lives. To become wealthy too early in one's life, however, leads some successful men and women to forsake progress and submerge themselves in debauchery and vice, a ridiculous type of personal failure. Gov. John A. Love of Colorado maintained that "moderation in pursuit of vice is no virtue."

Dignity comes to those who take pride in their work, and, certainly, pride is gained in dignified work. But neither pride nor dignity can be had by those who will not work. It is pride in himself that keeps man from becoming a vile creature. Voltaire believed that "Work keeps at bay three great evils: boredom, vice and need." Thomas Carlyle said, "Blessed is he who has found his work; let him ask no other blessings." Rudyard Kipling warned, however, "More men have died from overwork than the world justifies." A churlish friend of mine believed that more people die from worry than work because more people worry than work.

On a gravestone in Stephentown, New York, over the grave of Dr. Joshua Griggs, who died on January 6, 1913, in the 43rd year of his life, is this admonition:

> *Life's busy, anxious scenes are past, and here he finds repose. Sage sheltered from the stormy blasts of life's unblemished woes.Oh what availed his healing art, his skill divine to save? Death wing'd his never erring dart and sunk him to the grave.If worth departed claims a sorrowing tear. Pause friend or stranger, pause and drop it here.*

We are reminded daily that the road to the grave is short, that all must die. Let it not happen in the frenetic search for wealth through overwork. Achievements should occur without undue notice by the achiever. Anthony Trollope said, "In the fall of the leaf of a man's life, nothing can make him happy but congenial work today or the reflection that congenial work has been done."

If work is so great, why do so many people detest it? The Mexican's have a proverb which says "*Trabaje es sagrado; no toque lo*." Indeed it is sacred, but it must be more than touched.

My study

The Study

Just as a wise woman will have a room or a hidden corner for her activities in a conjugal home, so must a man have a study. It doesn't do for either to work on the kitchen or dining room table. Anyway, dining rooms are a bit passé, but if present, as they are in many homes, it is the "good" room where everything is exactly in its place, and you'd better not have dirt on your shoes if you enter. A man has to have a room where he can be as slovenly, disorganized, and unneat—or as neat, organized and clean—as he wants to be. To do the many things a man has to do to be a man, he must at times be alone. If I can't be in a building separate from the home, as my studies have been in the two homes we've built since 1949, it at least should have a door that can be closed, locked if necessary, to keep out kids and/or grandchildren. On the first study-office-library building I built in 1949, I put the door knobs up high within reach of a child 10 years old or older. This age of reason, I believed, would insure privacy without interruption while studying, writing, drafting, and dreaming.

Although rich men say they eschew desks, we rather more ordinary mortals who are scientists need at least two desks, a table, a drafting table, miscellaneous side tables for piling undone projects, books, reference papers in some kind of friendly disorder, plus lots of bookshelves and space for no fewer than six file cabinets. A couch would also be welcome. But a petroleum geologist has to have special files for completion cards, correlation cards, sample logs, geophysical logs, reports, and well completion reports. I solved that problem by having special files built by a master cabinet maker out of white pine and stained maple. Chairs must abound to seat clients, children who are in trouble, or people who are generally unwelcome, or on which to pile stuff when all else is overflowing. Of course, a complete computer system takes a different kind of organization for a study.

The modern family, if the man handles the finances, must set up financial and tax files, and keep handy account books, current correspondence files, and tickler files. If a woman handles the family finances, she must have the same, preferably a room of her own which may double as a sewing and weaving room. Lacking a room, at least a large space in a corner of the kitchen will do.

We built a second home in 1967. At first my study in the new house was the northwest room while I utilized my former study behind the large house. When I retired from teaching in 1976, we sold "Pueblo Buenaventura," thus precipitating a potential calamity. The northwest room was inadequate even though I kept a sizeable office at the university for academic research. Having had a double car garage built in 1954 on a single lot where our small house now stands, the solution was obvious: put a shop in one part and an office in the other. We built carports, a hobby port, a motorhome port, and a sailboat port, causing our friends to ask if we intended to roof over the entire 50 x 135 foot lot!

Guess what happened? Parkinson's law went into effect as my helpmate took over most of the house, except for our kingsize bed, part of the living room, and the kitchen. It was entirely logical that she have over half our small house for her writing desk, her projects, her looms, her genealogy files, her computer-printer desk, her boxes with our multi-thousands of color slides, her knick-knack shelves, her file cases, her sewing corner and an extra closet, plus a music corner with radio, cabinets, records, and tapes.

The point in this litany of a history of study-office-library is that such are necessary for both women and men. There is virtually no room for pictures on walls in our home, for they are covered with bookcases—as is my separate office, which has no telephone! It is amazing what one can do in an all-electric house with an electrically heated separate building. Many a marriage has been saved after retirement by having separate studies, by avoiding each other numbers of hours each day, but being within call and sleeping together in the same big bed, without snoring! We eat meals together, of course, and also watch TV together. Visits from children and their spouses and grandchildren are always welcome, but not usually in our studies.

Every family sets its own pace, honors its own time apart, and individuals arrange their own times to utilize that special solitude to get things done. Any family which doesn't plan for separate studies, even if only corners in separate rooms, is doomed to fail. So I say "Hail" to a large private study as a real modern necessity for active people who never really retire.

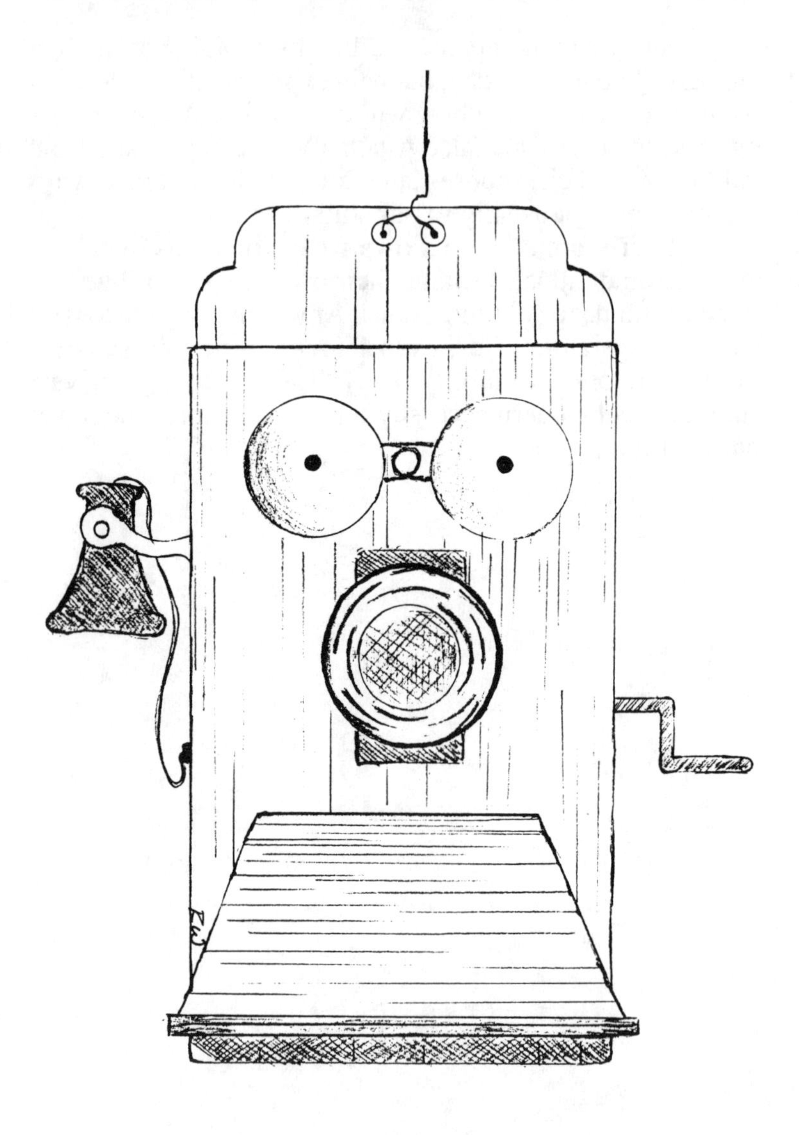

A sketch done in a telephone museum
of a phone used in the early third of this century

The Pesky Telephone

At times I wish Alexander Graham Bell had expended all his efforts towards inventing an airplane that could fly. He tried hard enough and failed. When he said "Come here, Watson; I want you" via an instrument made of wire, carbon flakes and a diaphragm that vibrated, he had invented a gadget of the Devil himself. What else can ring so raucously, jar one out of a pleasant reverie, jangle the nerves, and propel one out of chair or bed so violently as the jangle of a telephone? Not even a door bell has such power. And my father had two of them at our home in Ohio from two different telephone companies as early as almost anyone in the state of Ohio. He used one in his contracting business, but it rang, many times in tragedy, and was almost constantly used by neighbors who had no telephone and didn't know how lucky they were.

This fiendish contraption has now taken so many forms and is so portable that it has become a supposed necessity far beyond the inventor's original dream. And to what purpose? How can a person in heavy traffic loll back and talk on a cellular telephone while dodging tons of steel and rubber? I know of some philosophical folk who cherish their silence and individuality so highly that they won't have one in the house.

Telephonitis is a pernicious disease, even in my own family. When we had four children at home, our telephone, in a space of 17 teenage years, was constantly busy to the point where we would not even answer it during meal time.

Why this utter dependence on such a jarring piece of equipment? Some relief came about when we took our number out of the telephone book, but pressure from certain members of our family required the number to be available from information. This lack of a number in the book has confused many distant friends who have come through town and said, "Hey, why don't we call the Wengerds?" Not thinking about information, a very few think to call the department of geology at the university which gives out our number. You'd be amazed at the exasperation of some of our more dimwitted acquaintances who rail at us for not having our number in the book. Not long ago I got static from friends who said, "Why don't you get a telephone answering machine?" Clearly another satanic device! Why don't friends from afar think to write a note ahead of visiting? But an unlisted number, especially partially unlisted, is at the mercy of computer-generated telephoning. Such fiendish interruptions can only be answered by slamming the telephone back onto its cradle or lambasting the person who's trying to sell something or to tell you you've won some sleazy prize. Of course, that doesn't do any good, except for giving one momentary satisfaction. One couple I know won't answer the telephone until it has rung six times—a laudatory approach.

One of the greatest inventions involving this devilish apparatus is the shut-off, which we use frequently, especially during siesta time, frolic time and meal time. AT&T never allowed such a welcome respite on phones they installed, but that's been solved since the breakup of their hold on the telephone industry, which in every other way was a complete disaster financially.

Do I like the telephone? No! Do I use the telephone? Of course! But I don't enjoy using it, except when it is absolutely necessary. Despite my seething denunciation of this necessary invention, one must admit that the American telephone system is one of the wonders of the modern world, envied by many other countries, even though seldom copied. And, now, the greatest advance: telephone linkage all over the world by satellite. What will they think of next?

The Piling Theory of Filing

There are many ways to run a life. One is to have a place for everything and everything in its place. Why then, all things considered, does paper accumulate faster than one can find a logical place to put it? J. Paul Getty, one of the wealthy men of the world, was once asked what type of desk he favored. He said, "I don't have a desk! They get cluttered, so I let my assistants worry about desks." The solution to an uncluttered desk and an uncluttered mind is to have no desk at all. I have known many men noted for their perspicacity, wealth and self-discipline and have seen their desks, veritable piles of stuff and completely unorganized. In uncommon displays of intelligence, I've seen a number of them pick out on the first try a specific letter, document or newspaper article deep from within a pile of what looked like flotsam and jetsam. Such is the organized mind of a brilliant man. Challenged to tell how they do it, seeing the incredulity on my face when they pull that kind of magic, at least two or three of them have said to me, "You must know that a clean, well-organized desk is the sure sign of a small mind." Hurt to the core, but not showing it, I counter with the observation that these same men could never find what they wanted in a file cabinet. Also, I've noted that they—and I assume this applies to women executives as well—always have efficient secretaries who are warned to "touch nothing on my desk. If I can't find it, I don't need it when I want it."

In my university office, I had two desks, a table and a drafting table. In my consulting research office at home, I have three desks, two tables, plus side boards, a drafting table, and a light table, ALL piled high with projects, papers, letters, clippings, reprints of scientific articles, parts of three books I'm working on, various equipment, photographs, and only God knows what else. Although I know where practically everything is that I need at any moment to complete what must be done at any time, once something is finished and put into one of my six file cabinets, it's lost, almost forever. At my university laboratory office, there are four large boxes of unfiled stuff gathered during 30 years of teaching. At my home office there's a long work table in my shop, under which are six giant cardboard cartons of papers, clippings, letters, and photographs gathered since we moved to our present home in 1968. I can't really remember what's in all those boxes!

Mark Twain said he had a fool-proof way of answering his voluminous mail. He'd put letters he received into a drawer for six months, and then he threw them out as it would be unnecessary to answer them. Dr. Paul Weaver, a former president of the American Association of Petroleum Geologists, went Mark Twain one better. He'd read letters he received and promptly throw them away, unanswered!

How, then, did the ten boxes of unfiled stuff mentioned above come to be? By my incomparable theory of "filing by piling," things that don't have to be done right away are piled in three piles: "Do Tomorrow," "Do Next Week," and "Do If I Ever Get Time." This reminds me of a file I once saw in another geologist's office labelled "Stuff I don't know where the heck to put."

Obviously the technique of piling stuff one cuts from magazines and newspapers requires that the piles be filed in some system if one ever expects to use what one has read in a career of writing and lecturing. Many items should be filed under two or more headings, necessitating the onerous chore of copying or the hiring of 32 secretaries and researchers—as one famous author does—to write thick tomes on the subjects and areas he chooses to write about.

This piling technique follows Mark Twain's simple system in one way: every once in a while, a pile is gone through, and the things one should have done last year are answered with due apologies. The rest are piled onto an older pile of things undone, and finally, a year or so later, when the pile is about to drive one out of the office, you throw it away. I have started fires in the fireplace with anything that will burn. If it won't burn, put it between the rafters of the ceiling of the shop as insulation when you replace the ceiling. Or do as I have now done since 1947. Put the piles into ten boxes and stow them in newly built work spaces or into a garage, banishing your car to the elements. Then hope you live long enough to examine them sometime!

People better organized than I am, men of great creative ability, delegate others to handling chores they find onerous. But to whom, if you are an independent, does one turn to organize these piles of stuff, these reams of paper, these thousands of clippings? Hire secretaries? If you have two or more, you're working for them, and your days of contemplation are over. Press your wife, children or grandchildren into the fray? No better way to be totally unpopular. There seems to be only one sure solution: die and let your heirs throw the stuff out!

At my desk at home
1993

Preparing a bottle for throwing overboard

The Pack Rat

There is something marvelous about a pack rat, and I'm talking about that rodent with the unlikely bushy plume on his tail. He (including a she) hides in old mines, shallow caves, ranch barns, tack storage areas, in fact any place mainly in the West. Usually they work near human habitation where they can gather glittering objects or almost anything they can carry, and you'd be amazed at what a pack rat can accumulate. I've seen them and their thousands of valuables under highway overpasses, under corn cribs, in abandoned houses in ghost towns, in mine tunnels between lagging, and in aged woodpiles, but the places they live must be dry. The selections of what they steal would boggle the mind of any avid collector: coins, rings, bottle tops, beer can pull tabs, pencils, springs, sticks, old teeth of other animals. I've even seen a set of false teeth in one nest, and I'd venture the owner never thought to search in a pack rat's nest. Their nests are distinctly unneat. They can't take time to be housekeepers or to keep their treasure troves in any semblance of order. So pack rats are sloppy, just like their human counterparts.

When you find a pack rat's nest, for they're easy to find where sticks and grass tumble out of wide-mouthed openings, you'll need a flashlight to explore the many nooks and crannies filled with precious acquisitions. Also, the flashlight will help you note if the occupant is at home, for the pack rat has very sharp teeth and large, beady, black eyes that shine; you'd swear they shine in the dark. Don't, under any circumstances, reach into a pack rat's nest without gloves because those critters will bite to protect their collections. And their bite is deep and painful, though not usually dangerous. Why? The mouth of a pack rat is cleaner than a whistle—certainly cleaner than a human's befouled mouth.

Though they do not live in colonies, there are always two, with the female being the chief acquisitor. What else? She's a female, typical of the gender! I don't know of any rodentologist who has made a deep study of pack rats, but I'll research that with my friend, Dr. Findley of the biology department at the university. He's the West's expert on small mammals. Many people send him specimens of odd little beasts, and he collects thousands himself. They say his laboratory is the rattiest one on campus. I hope he realizes I say that as a joke!

There is one very interesting thing about pack rats that seems to happen among no other individuals of any animal group. When they swipe a collectible, they always leave an object, usually not as valuable as what they take. Also, pack rats are not only cute, but they have wise-looking, quizzical faces; they look like they're always trying to find out what you might have that they can purloin.

The Inveterate Collector

In the spectrum of humankind, there exist individuals who have acquisitive propensities difficult to fathom. Makes one wonder about oneself. We observe people who collect matchbook covers, stamps of foreign countries, barbed wire, bottle caps, coins, old tractors, models of cars, old gold watches, glass dishes, paper clips, smoking pipes, old fountain pens, in fact anything regardless of value. These people call themselves collectors, and for every object that such pack rats collect, there is a local, regional, or nationalclub for them. Interestingly, the herd instinct of collectors leads to club constitutions, officers, dues, formal meetings, speeches, rules of order, bylaws, and newsletters—a typical bureaucracy.

There seems to be no reason behind what people gather to trade, sell, keep, and gloat over. Whether it be stamps, coins or baseball cards, they spend hours arranging, displaying, fondling, classifying, and rearranging these objects with glee. It keeps them off the streets and out of taverns. Each person to his own bucket of blood, regardless of value or meaning, but there is one type of collector that gathers up things that everyone else throws away: the inveterate pack rat. It is a syndrome, a psychological hang-up, and can be very costly of space, not to mention time. No clubs are involved, and there is no apparent value; some of these objects are so common as to beggar concern.

Consider the tin can. One thinks of tin cans as negligible objects which might be saved because there are so many types, sizes, designs, and colors. Think about the revolution the lined tin can has brought about in the distribution of fruit, vegetables, meats, beer, very nearly anything that needs to be preserved and shipped by truck, boat, railroad, or aircraft. I save tin cans, not for the above considerations and not indiscriminately, but because I like them and find them useful in first, second and third derivative ways. Suppose tin cans and aluminum cans should suddenly cease being made. Very quickly they'd become objects valuable to collect, to classify, and to trade, and their value would rise as new collectors would see them as objects that represent a segment of human history.

For some 43 years, I smoked pipes and bought one pound tins of Heine's, Bond Street, Edgeworth, Prince Albert, and Revelation pipe tobaccos. Throw those lovely cans away with no regard for their collectability? Some years ago I suddenly realized that the old one pound tobacco tins became 14 ounces, and then 12 ounce tins, yet the prices advanced astronomically. But something else happened. They came equipped with snap-brim lids which had a rubber seal. Aha, a use sprang to mind. One could store stuff in such marvels of engineering, and over the last 20 years of my pipe-smoking days, I saved them. They cluttered up my garage and filled shelves in my workshop. What to do? Today at least 50 of those clever cans are filled with pinto beans, which last forever and are one of God's gifts to the human race, especially for us chile bellies who love boiled pinto beans. Clean, dry pinto beans, in sealed tobacco tins, line the east wall of my motorhome port. I'm the bean maker in my family, and that food is a staple in our diet. How I prepare them is a secret. A minor measure of what little fame has come my way can be attributed to my bean-making ability!

I've often wondered if there is anyone who collects oil cans? I don't, but it is intriguing to consider the evolution of the one quart oil can. I can remember when there wasn't any such can. Oil came in bulk. You pumped it out into a tin pitcher and put the oil into your car. Suddenly, stout quart tins of oil were available. Then the oil companies, in their vast marketing stupidity, started putting oil in paper "cans." Now, more logically, oil comes in easy-pour cans made of plastic, what else but a petroleum product—smart move. But the cans are not really collectable. Yet throwing them out wrenches one's pack rat soul.

My favorite is still the tin can. They nest so nicely, come in so many sizes and shapes: ribbed, unribbed, striated, plain, with ranges of zinc coatings. Why can't they be made square so they fit better in portable ice boxes, electric refrigerators and on storage shelves? A sad lack of design finesse. I do not collect them, but I keep as many as I have room for because they house nails, bolts, staples, a thousand and one bits of metal which will someday, no doubt, be useful sometime for something. I use them to clean paint brushes in solvent and to store leftover paint. I cut the bottoms out and flatten them to cover knots that have fallen out of our wooden fence. My wife maintains I keep them because I'm afraid they'll quit making tin cans. Not so. I keep them because I think they're a great invention and handy.

There are other pack rat items that grace a large space in our various storage areas: glass jars with lids that seal. The reader may think this is some kind of mania, but I have reason to believe the glass jar may not be long with us in view of our penchant for using plastic. Take for example that abomination, the plastic tumbler. It's downright sinful to have a tall cool drink out of plastic when glass is such a friendly substance for housing a cocktail. Who can imagine our mothers and grandmothers canning vegetables in anything but glass jars? Moonshine would be tasteless if one bought it in plastic containers in the hills of Virginia.

Glass bottle collecting is one of the truly great hobbies of those who rummage around in ghost towns. Glass that turns violet, blue, and magenta as it lies for decades in the hot desert sun has great value. But modern glass is now so pure that most of it has no complex manganese-cupric impurities to color these modern glass bottles. Another glass hobby involves glass insulators. When my buddies and I were boys with our first rifles, we'd shoot them off the tops of telephone poles, not realizing that we were ruining potential antiques. Of course, we got caught, to more than our consternation. Today, a glass insulator collection can be valuable enough to pay for a college education.

There's another kind of pack rat who remembers what happened to silver. A Spanish friend of mine from New Mexico has buckets and tubs filled with copper pennies. I haven't the heart to tell him that most modern pennies are made with zinc and contain only enough copper to cover the zinc. Not surprisingly, the bright, shiny zinc pennies of World War II are now truly valuable collectors' items, and I have two of them given to me by a Cajun friend with the unlikely name of Robby Burns. The lack of foresight on my part involving silver is demonstrated thus: From about 1950 to 1960, I saved silver half dollars—1200 of them—by tossing them in a corner between abutting bookcases. After taking my family to Europe in the summer of 1960, I felt financially distressed, carted the half dollars to the bank, and deposited the $600 swag into my checking account. Today that hoard of silver half dollars would be worth about $3000, even in a worn circulated condition. The modern nickel-tin, crappy half dollar, now going out of style, is valueless as metal and almost valueless as money. So who knows, tin cans and glass ars may become valuable, but don't bank on it. Their value lies in present and future usefulness!

Ready to go to work
Seymour, Texas
1941

Throwing a bottle
from the deck of a freighter
1987

Part Six

MONEY

Money and Wealth
Any Idiot Can Spend Money
America—Watch Out!
Seven Swings at a Million

Money and Wealth

It is true that if you don't have at least some money—called "walking around money"—life can be most difficult. As much has been written and spoken about money as about sex or women or any other subject. There is an old Jewish saying by a man named Bondi that "It's much better to be rich and healthy because a poor person gets no benefits by being ill." In fact, poverty is not only unhandy, but it's also darned embarrassing. In the Alyeska office of the company set up to hunt for oil on Naval Petroleum Reserve No. 4 in Alaska was a desk sign that read "We live by the Golden Rule. The guy with the gold makes the rules." This brings up advice my father gave me with the first nickel I got as a four-year-old boy: "Always have money!" Much later, when I could really understand his advice, he told me, "A poor man can't do a lot of good for other people, because the world we know runs on money." Father's mother always chastised him severely for "never thinking about anything but money." Ironically, she herself never had any because as soon as she got some, she spent it.

Lloyd Shearer, that canny editor of "Parade," a supplement in many Sunday newspapers, had it right when he said "It has never been sinful in the American ethic to make a buck." Of course some people can't handle money properly, which led my friend Art Mosher to observe, "Some people never make enough money to make asses of themselves—and that's good!" A family I know—rather more closely related to me than I care to admit—once told me, "We have never lived within our income." As a consequence their offspring say, "We don't like to handle money. It's too complicated." Frankly, it takes very little ability to tell oneself "Any idiot can spend money" and then follow Ben Franklin's advice that

"A penny saved is a penny earned." Gandhi observed that "Poverty is the worst form of terrorism." Including myself, there are many people who have been broke but never poor. In most instances poverty is a result of money mismanagement. Many young people think money burns holes in pockets, and so they get rid of it. If you make your goals more important than your temptations, who can stop you from acquiring money? Sir Peter Scott said of the rich, "The uppercrust is a lot of crumbs held together by dough." Jonathan Swift set the tone with this couplet:

"Get money, money still
And let virtue follow, if she will."

Honoré de Balzac, the great thinker who was oft misquoted about women and had much to say about many things, made one outstanding comment that I've noted as a truth: "Rich people will never be capable of understanding the difficulties of the less fortunate." I've known a great number of wealthy people. Most do not try to understand poverty, do not believe in bad luck, but do try to help those truly in need through no fault of their own. How does one determine whether an unfortunate person or family really is deserving of help? You can't help unfortunate people without knowing the facts, and precious few will tell you enough to enable you to determine the causes of their ill-fortune. In every case that I personally know about, full facts either reveal lack of financial discipline or else blame so-called "acts of God." One should help the latter and not give financial aid to the former. If it is not too late, the answer for most people is to provide education and opportunity so they can learn to help themselves.

Uncle Remus said, "Money ain't no good unless you ain't got any." But in a reverse kind of logic involving wisdom, I've found that some people appear to have gained the status of having wisdom because they didn't make enough money or save enough money to make fools of themselves.

Money yields freedom if you dominate it, rather than letting it dominate you. It is evident to any intelligent man that you can't make up with money what you lack in judgment. Why? Without good judgment money disappears—rather quickly!

Much is said about seeking advice on how to become wealthy by investing money. This involves hiring financial advisors whom one pays directly for advice. Brokers who sell or facilitate buying stocks, bonds, financial paper, and real estate are always suspect, because commissions are their bread and butter. Edgar R. Fiedler noted, "Forecasting is very difficult, especially if it's about the future...Ask five economists for an opinion and you'll get five opinions—six if one went to Harvard." Colonel G.D. Ashley mentioned that "Skill in handling numbers (hence money) is a talent, not evidence of Divine Guidance." There is the old saw that "It takes money to make money," which is trite but true! And how does one get money? By judicious saving and avoiding luxuries one doesn't need. In investments it is good to listen to Burton Hillis: "We might be more eager to accept sound advice if it did not so often interfere with our plans." In investments, as in religion—if God be with us, who can be against us—but when it comes to money, it's each man for himself. Alfred P. Sloan, in a rare bit of philosophy involving money and investment, said, "I've made many mistakes, but one mistake I never made was to make things too small."

The youth of today do not seem to realize that one can't make money by investing in things that decrease in value through time. Andrew Carnegie, that savvy Scot who started Bethlehem Steel, said, "My definition of success is this: the power with which to acquire whatever one demands without violating the rights of others." It has always been true that to the shrewd belong the spoils. Opportunities in money-making require risk—the greater the risk, the greater the potential return. Don Quixote thought that reason is a veritable monster and that wise madness is better than foolish sanity. Those are the sentiments of a plunger—and we all know what happened to Don Quixote according to Cervantes. Yet plunging is how some fortunes are made. Each man who invests should reserve part of his risk-taking money for plunging— wrong-headed as that advice may sound. Circumstances are indeed the rulers of the weak and the opportunities of the wise. Robert Browning, in "Andrea del Sarto," put it another way: "Ah, but a man's reach should exceed his grasp, or what's a heaven for?" C. Wright Mills, writing in that excellent book, *The Power Elite*, stated the following: "Of all the possible values of human society, one and one only is the truly sovereign, truly unusual, truly sound, truly and completely acceptable goal of man in America. That goal is money and let there be no sour grapes about it from losers." That's quite a raw statement! De Tocqueville considered that the love of wealth is at the bottom of all that Americans do, and that shocked this most astute observer of America. Every one knows that "Money is the mother's milk of politics,"yet G. K. Chesterton averred that "The golden age comes only to those who have forgotten gold." Either he had enough, or he gave up trying, or he believed in a moneyless system which has never really worked anywhere.

A money society is naturally competitive. For some to win big, many must lose some. Once it is learned that your dollar is fair game, that your savings are not secure, and that what you have is rather easily put in danger through lawsuits, fair as they may be, the entire regimen becomes one of beating the system. The Spanish have a proverb which says that "Living well is the best revenge." In a society of inflation—which occurs whenever there is a taxation system designed to take from those who have to give to those who have not—it is well to do two main things: live well, but within your income, on money made from investments in situations that beat inflation. Possession of capital is not enough; it must be made to grow faster than inflation plus taxation. This challenge is one of the greatest games the middle class plays in America. The much publicized "standard of living" concept put forth in advertising is a false concept. Americans live too high on the hog, and, as long as most of them do, there are enormous opportunities for making money by beating the system.

Anyone who can read, write, do mathematics, and handle a saw, hammer and drill can build his own house. It doesn't take a Ph.D. to read plans, design your own house and build it. Certainly it's hard work; contractors hate to see you do it, and friends marvel at the guts it takes to beat the system. What made America great was the can-do spirit. Not to try is to fail. Rockwell Kent, a great artist, answered a question about his philosophy as follows: "Do you want my life in a nutshell? It's this: 'I have only one life, and I'm going to live it as nearly as possible as I want to live it.'" The idea that fate controls our progress is false. That we must keep up with the Joneses or lose face is ridiculous. All people have problems, and most people lead lives in fear. Mark Twain said, "I've lived a long life, and I've had many problems, most of which never happened." Satchel Paige, that philosophical black baseball player who was put into the Hall of Fame late in life, said, "Never look back; something may be gaining on you." Ralph Aling, Ohio's own genial home-spun philosopher, now 96 years old, said it most simply: "Success does not crown the average man in life—only the unusual." With average good health, a modicum of smarts, enthusiasm for living, and a willing, supportive helpmate, success can be the lot of any American unafraid to try!

Samuel Johnson mentioned the insolence of wealth and the arrogance of power. Why do so many people equate money with power and wisdom? One of the most powerful men ever to live was broke all the time; he was never in poverty because he had friends, but, being a revolutionary, he also had many enemies. He kept upsetting the stolid status quo. He advised his followers: you must be absolutely fearless, absolutely happy and constantly in trouble. His name was Jesus. It seems to me an active investor could well take those three admonitions to heart. H. Ross Perot, one of the most successful investors in America, said, "If you go through life worrying about all the bad things that can happen, you soon convince yourself that it's best to do nothing at all." Sören Aabye Kierkegaard believed that "To venture causes anxiety; not to venture is to lose oneself."

Robert Louis Stevenson thought that "It's not by any means certain that a man's business is the most important thing to do." However, deep thought and inaction are often thought to be a sign of lassitude. In fact, we Americans often rely on failure to excuse our laziness. Henry Ford was once twitted for keeping on his payroll a man who sat around apparently doing nothing. Ford retorted that the man was his most valuable employee, and he paid him the highest possible salary because he came up with valuable new ideas on how to make more than just better cars. The man's name was Wilbur Stout, inventor of the Ford trimotor airplane and the Fordson farm tractor!

Finally, there are so many pithy comments made by so many famous people that their triteness should signal general acceptance and action by everyone. But such is not the case. I hasten to remind you of Mark Twain's toast at a banquet in his honor: "Here's to the fools. Were they fewer in number our times would be more difficult!" It was J. Paul Getty, at one time considered to be the richest man in the world, who said, "A man who knows what he is worth is not worth much." During investigations into the silver scandal in 1980, Bunker Hunt said almost exactly the same thing before a congressional committee: "I do know people who know how much they are worth; generally they aren't worth very much." This is akin to J. Pierpont Morgan's statement: "If you have to ask what a yacht is worth, you can't afford it." Must one conclude that outrageously wealthy people have the same attitude toward those of us only moderately well-off? These are harsh and unfeeling judgments of money as a criterion of the human condition. Certainly there are more worthy attributes of an individual's value to society, but it is a truism that a man in poverty cannot be as helpful as can a man with money.

Any Idiot Can Spend Money

There is no less a deserving soul than he who has unbridled desires and a strictly limited income; or, on a plebeian note, it is said of some people that they have "champagne tastes and a beer income." To do without may be humility, but to be a slave to things borders on stupidity. To go into debt to satisfy one's desires for other than necessities seems impractical. How about gearing one's purchases to one's ability to pay cash rather than go into debt? There is an old New England saying that goes "If you can't pay for it, you don't need it." My father once told me credit was what made the financial world go round—the best thing that ever happened involving money—yet he counselled that one should only buy things that increased in value through time, except for necessities that are used up while one stays alive in comfort.

So there you have it: if you can't pay cash, don't buy it, and if you need it to make more money, go into debt but get out of debt as quickly as possible. It was Ben Franklin who said, "A fool and his money are soon parted." As to scale, "penny wise and pound foolish" was an ever-present phrase in early Pilgrim homes. And the Pennsylvania Dutch caution one never to use up money like it's going out of style. One of my spendthrift friends believes he should spend it before it burns a hole in his pocket, and pockets are difficult to repair. The spending of money should never be substituted for the application of logic.

Inflation ruins savings that earn less than the inflation rate. Cars, airplanes and boats depreciate, whereas homes and properties generally appreciate in value at rates greater than taxes can eat up their increase in value. However, taxes, inflation and living high on the hog have ruined more salaried people's budgets than any other combination of fiscal conditions. The financial careers of monied men and women progress from creative production to imaginative investment, to jealous preservation of capital and to wise spending during an entire life. Of course splurge if you can afford it. One of the best pieces of advice I ever got came in two parts: "Never gamble with a destitute man," and "Never bet on anything that can talk." Sure, go ahead and gamble on situations that bring highly productive returns; if you lose, at least you've tried. Reserve enough to live comfortably, and be sure to give weekly to charity because every dollar you give comes back tenfold. You can't do God's work from a position of poverty.

The consumer revolt depends on the proper timing of the use of two important declarations: "I'll be darned if I'll pay that much!" and "No thanks, I don't want to buy it." Enough people raging against high prices can only result in lower prices, and if that results in no further production of needed items or services, make the item yourself, or do the service yourself, or find substitutes. One of the glaring untouted untruths bandied about by unscrupulous people is that to do without lowers one's standard of living. He who controls his emotions controls his world. When we were married in the perilous time of 1940, we declared, "We will be two against the world; we will win over those people or situations that attempt to take our hard-earned dollars and investments." That declaration has paid off handsomely. We have never borrowed money to buy a car nor to build a house. We saved to pay cash in order to avoid interest. We built our own houses as well. All it takes is a hammer, saw, trowel, and common sense. The sanest advice I've read recently was a mother's advice to her spendthrift daughter: "Any idiot can spend money."

America—Watch Out!

There are false attitudes rampant in America—attitudes that make a hollow shell of a country where freedom, liberty and opportunity have always been considered the avenues to success for a thinking person. Why have we become so enamored with acquiring money that we are losing our ability to make things or create new ideas with our brains? Why is it that we Americans think that the rich are always right because they are rich? This is an especially great failing when we value rich spendthrifts, people who inherited money rather than having earned it through creation of things and ideas that bolster everyone to a better life. Why do so many cater to the well-dressed and treat shabbily the intelligent who create books and philosophies that make the world a better place? Elegance costs much more money than it's worth!

It's a real mystery to me why a people who eschewed the rigid class structure of Great Britain are today so intrigued by the British accent and the royal monarchy. There is no doubt that the calibre of a man can be measured by what impresses him or by the intellectual level of people he tries to impress. Of what earthly good is a fancy limousine when compared to any vehicle that gets you where you want to go? Or a house so gigantic and ornate with a mortgage so large that one's entire effort must be directed toward making huge monthly payments? Important as credit is for obtaining the means to create, why do people go so heavily in debt to impress other people?

Why do people dye their hair? Because they're afraid of age? A critic of mine—a man deathly afraid of dying, a man so wealthy he couldn't spend all his money if that's all he did for the rest of his life—twitted me recently by noting that my hair was getting whiter every year. He thought to insult, but my response shocked him: "I'm proud of my gray hair. I've earned every last one of them in my service as a professor in a university." He and a number of my wealthy friends are convinced of my apparent cupidity in not bowing down to their wealth. They may be smart, they may consider themselves to be self-made men, but in intellectual discussions of a truly basic philosophy of living, they rank down so far as to be unworthy of comment.

A T-shirt slogan that I noted once seems to strike the level of many people's thoughts about what's important: "The successful man is the one who dies with the most toys." Why do so many people die with so many of their toys unpaid for? It is not unusual for those who have few material possessions to be mildly upset by the obvious profligate waste by those who have many possessions. Yet pride in mental prowess completely undirected toward making any money is rather a false pride when money is so easily earned with about ten percent of one's brain at work. Why not spend the other 90 percent of one's time in doing good works? One of the great tragedies involves people who make vast sums of money when they are young or who inherit valuable estates created by capable ancestors. Menander put it thus: "No just man ever became rich all at once." I have a number of acquaintances who made financial killings before the age of 45 and have been unhappily frittering away their lives, time and health ever since. Remarkably, they show signs of intelligence but lack all ambition to help anyone in any way. How can that be?

If people with lots of money really think that worshipping money for what it will buy will allow them to live longer, I have news for them. It won't! If they can't think what to do with it other than buy what's not necessary, they should set up foundations and go sit in a corner and contemplate their futures. They'd live much longer. Also, they might jot down a few thoughts as hunger mounts. Who knows? Some may even become prophets in the wilderness, railing at profligate government, big business, the possible falsities in universities, and the unbridled but benign embezzlement that occurs because of the scramble for money, especially by banks and certain companies.

If musing is contemplation by the soul and dreams are the musings of one's soul, why not sit back and do a bit of musing about how America got into this mess? Consider the national debt, the balance of payments debacle, the credit crunch, the options market, the silver fiasco, why so many people save so little money, why feather-bedding is allowed. Why do we keep electing a Congress so bent on ruining America with spend-and-tax as their chief obsession instead of representing all Americans? Must we forever be saddled with senators and representatives who try to spend America into supposed greatness based on money? Why do we not pay attention to those cries in the wilderness? How can we allow so much crime, the destruction of so many jobs by mergers and acquisitions of companies, so many homeless people, such vast overruns in defense contracts, so many crooks in banking and elected positions, so much thievery on our docks, such immoral rises in the cost of mailing letters, and so little effort to construct adequate housing that young people can afford? Why all this, America?

America has developed a false house of cards based solely on money and its acquisition. Have we lost our souls as a country? Just who are the devils in the wood pile? Or did I just name them above? Is it revolutionary to suggest that everyone should have an opportunity to work, that adequate pay should result for anyone who will work hard? Was Pogo truly right when he said, "We have met the enemy and they are us"? No leader of a moribund people can right things that are wrong in a democracy. The correction must come from within every citizen of this great country, not from above. Jefferson was right when he called government a beast. Governments produce nothing, but people do, and confiscatory taxes will be our ruination.

America made a number of mistakes whose results have come to roost in our body politic. The first was the spoils system under President Andrew Jackson. The second was Abraham Lincoln's statement: "Government of the people, by the people and for the people." A government should never be governed "by the people" but by representatives elected by the people. The third great error was the direct election of senators by the people instead of by representatives elected by the people. The fourth was the institution of the federal income tax in 1913, a grievous error which resulted in giant government far beyond the power envisioned by the wealthy men who signed the Declaration of Independence.

There is a practical solution based on who should do the taxing and on what is possible. Think about the problem and come up with some suggestions! Sufficient to say our America is a unique country, the only place in the world for independent people to flourish in mind, spirit and body. We should work hard together and quit worshipping money for money's sake.

Seven Swings at a Million

The soul of learning is desire. The mechanism involves repetition as well as the use of whatever intellect is available. A prime example was Stephen Widfield, who graduated from Stanford University with a doctoral degree in geology just before World War II. He accepted a teaching job right out of graduate school as an assistant professor at Earlham College in Indiana. After one year, he found himself dissatisfied and went to work at the Map Construction Division of the U.S. Army Air Corps Center in Washington, D.C. While there, he was sent to Reno, Nevada, to research a site for a major airfield in the wind-shadow of the Sierra Nevada. Some long-haired savant from Yale had suggested that the training of fliers in a locale that had known williwaw-type winds cascading off high snow-covered peaks would be great preparation for Alaskan duty. It was expected that the Soviets or Nazi Germany or Japan would make an attempt to attack North America via the frozen north of Alaska.

Steve married before World War II began and took a vacation the summer before he went into teaching. In Reno he became acquainted with a lawyer, the father of one of his close friends at Stanford. Driving over the Mount Rose Highway, he noticed eastward-sloping, sage-covered range land north of the highway with a snow-fed stream gurgling along. Although he had signed the contract to teach at Earlham, he nonetheless always wanted to live in the West. So he dropped in to see the lawyer, a kindly old man, and asked that he research ownership of the land southwest of Reno. When he returned from his vacation, he found a letter from the lawyer. Yes, the land was available, a whole square mile at three dollars an acre plus closing costs, a small stipend for the lawyer, and costs of a re-survey of the property. The total cost: perhaps $2400. But Steve was $6000 in debt for his graduate work, had just married, and had a new contract to teach at $2000 per nine months. How to finance such a purchase? He knew the acreage would some day be valuable, so he contacted his father who lived back in Indiana where Steve was born. His father was a doctor, and, as everyone knows, doctors are notable failures as investors. "No way will I lend you $2400 to buy land in Nevada. Who the heck wants to live in Nevada?"

Steve had no credit rating whatsoever, so banks would not lend him money. He hadn't married a rich girl, as his father had advised him to do. All efforts failed, for who would want to lend money to a young assistant professor of geology?

Twenty years later, World War II a distant memory, Steve and his wife, now with three children growing up, drove along the Mount Rose Highway, and the square mile of prime range land was covered with large houses. His first swing at becoming a millionaire had failed.

Steve, whose father was moderately well-to-do, had had to work his way through Stanford because his father wanted him to go to the University of Chicago and become a doctor. Headstrong and independent, he also crossed swords with his father about joining the Army Air Corps. Because he had spent a year at the Army Map Service, he decided to become an Army pilot. After training at Randolph Field, he was sent to the new airfield at the site he had a hand in choosing. Because he was a superior pilot, the Army Air Corps, in its usual peculiar way, next assigned him to duty as an instructor at an airfield in California. His family was starting to grow, and he couldn't save much money, despite his now increased salary. He saw great opportunities in buying coastal property south of San Fransisco, did some analyses, both financial and by demographic projection, but to no avail. "Wartime," said the banks, "was no time to lend money to an Army Air Corps pilot to buy land." The second swing at becoming a millionaire passed by.

The Wengerds—
Tim, Diana, Sherman, Stephanie, Florence,
and Anne
1959

"Why this penchant for becoming a millionaire anyway?" pondered his wife. Steve had been born during World War I and grew up and went through high school and college during the Great Depression. Though never poor himself, he knew what it was like to be broke, and he could see the results of money mismanagement as well as the terrorism of hunger among thousands of people in bread and soup lines in the Midwest and on the West Coast between 1929 and 1939. He was determined to become a millionaire!

After two years of instructing Army Air Corps pilots and following the birth of his second daughter, he asked to be sent overseas with a newly formed Air Corps fighter squadron. Not knowing whether they would head to Alaska, the Pacific theater, or the European theater, the squadron trained for Arctic duty, desert duty, and possible duty in the tropics. Again, like the vagaries of duty overseas, much training was apparently wasted because his squadron was sent to Guam after its capture by the U.S. Marines and the U.S. Navy. As fighter pilots, the squadron flew missions on the first legs of B-29 flights over the Bonine Islands south of Japan. When the atom bombs were dropped on Japan, the war came to an abrupt end, four years sooner than anyone expected. By now, Steve was a colonel in the U.S. Air Force. Gen. Hap Arnold had changed the Army Air Corps, which had been in the Signal Corps, to the U.S. Air Force. Steve had now saved enough money so he could pick and choose. As a graduate geologist he decided to go to work for an oil company rather than stay in the Air Force at reduced rank.

Back in the United States, he found a position with a major oil company in the Gulf Coast region and stayed with them for five years. He learned all he could about oil exploration, production, acreage leasing, and well-completion techniques. Suddenly, he was offered teaching jobs at both the University of Nevada and the University of Colorado—a fork in the road! Although he had taken some economic and mining geology at Stanford, his interests lay in teaching petroleum exploration and in becoming a consultant. So he took the position of Associate Professor of Geology at the University of Colorado. Steve's father did give him some good advice, such as: "You can't become wealthy by working for others"; "You can't make a million by working just eight hours a day"; "Savings won't do it. You have to invest in things that will increase in value without your efforts"; and "Take some money and plunge." While he was working for the major oil company, his genial boss, a vice president, had told him he could become well-fixed if he saved half his raises plus ten percent of his net salary plus ten percent the company would put into a stock fund. This latter ten percent the company would automatically withhold and invest for him in a company-controlled international fund. "Whew," thought Steve, "what will my wife say to ten percent out of our salary and an added ten percent of savings on our own?" But his wife was a game character. After all, hadn't he been overseas for two years? Hadn't they even saved about 20 percent of his Air Force pay? And to boot, the company was investing another ten percent for his account. Why not give it a swing? A petroleum geologist needs to have a great wife if he is ever to amount to anything!

Going up the Grand Canyon on the Kaibab Trail
April 27, 1971

When Steve and his wife went to Colorado, their oil company fund in both stock and the International Fund totalled $16,000, a tidy sum in 1950, especially considering that Steve had stayed with the company just a little over five years. Within a month of his move to Boulder—where the university furnished low-rent housing and a salary of just half of what he had been earning with the oil company—his former boss called and said, "Our company is not represented in the Rocky Mountains. Would you accept a consulting position?" Of course the answer was yes, and Steve began assembling files, published literature, drilling data, weekly scouting reports, and monthly exploration reports on special projects in his spare time on evenings, weekends and during summer vacations. Over the next five years, he earned annual consulting incomes between $12,000 and $30,000, and he began serious investment programs. His teaching income in the first five years rose from the associate professor salary of $4500 per year to $15,000 per year. The hordes of veteran students, funded by the G.I. Bill, had graduated. By 1955 his consulting association with his former oil company had resulted in the company opening three exploration offices in Casper, Denver, and Albuquerque, and his services were no longer needed. For five years, with the help of people he had hired—draftsmen, geologists, engineers, and landsmen—Steve had found a major gas field, two large oil fields, and had helped the company set up the three division offices in the Rocky Mountains. He had worked himself completely out of a job, and, worst of all, he wasn't allowed to take royalties or working interests on what he had discovered.

What to do? Staying at the university was not a dead end. When he became a full professor, his salary still would be only about $40,000 per year at a maximum. If he became a dean, a vice president or president of the university—three jobs he had already been offered at other universities—his salary could go up to as high as $90,000 per year. Detesting administrative work, he avoided that route and remained a full professor, a position he considered to be the highest rank in a university! His estate now had a value of some $200,000, still a long way from a million dollars.

Then came numerous other consulting jobs where he could earn or buy royalties, overrides and working interests. He had not been allowed to do this while working for his former oil company as either a salaried employee or as a consultant, so his efforts to hit the million dollar mark as a company man had failed.

His swings at becoming a millionaire through investments were thwarted by recessions, a few bad investments, and an unsuccessful plunge or two, but he gathered up a hundred thousand or so. So he considered that route a failure in his efforts to become a millionaire.

Steve and his wife—all through their civilian earning

period to the time he could, as a consultant, acquire interest in exploration ventures—had always allocated ten to 15 percent of their income to charity. They helped support their Methodist Church, their colleges (even Earlham where Steve taught for only a year), the Salvation Army, a mission in Guatemala, Stanford University, and a host of lesser charitable activities. They also travelled extensively so their children could see the world. By this time, becoming a millionaire wasn't as overriding an ambition as living a life full of experiences. He now taught and did oil exploration work because it was fun.

One day a local entrepreneur in Boulder called Steve and said, "I have a group of people interested in putting together a fund to find oil in the Denver-Julesburg Basin east of Denver. Would you be interested in selecting a block of acreage we could acquire?" This sounded like fun because he had looked at a favorite part of the basin for his old boss but was turned down because it promised too small a return for a large company. Following the ethical code of his petroleum exploration society, he asked his ex-boss for permission to use the information he had assembled. He received a letter of permission and reported to the entrepreneur, "O.K. I think we can make a play for a shallow oil field. If you pay the expenses of final exploration by surface mapping and photogeology and give me an interest, I'll work it out and write a report with maps to show you the area and the costs involved for road-building and drilling plus completion as an oil producer or, if a failure, as a dry hole according to state and federal regulations."

Thus was born the Golden Horn Oil Company with nine investors, in which Steve was to receive costs for exploration and well-site geology while the well, or wells, were being drilled. The entrepreneur put together the funds and agreed to assemble the acreage comprising state lands, some federal lands, and fee lands owned by a railroad. He gave Steve a carried ten percent interest in the venture. After three months of spare time work, even doing some of the surface plane table mapping during a snow storm at

Christmas time, the report was finished. Initial work on aerial photographs had shown the existence of a fault crossing the plunging nose of an anticline. The acreage was assembled, location staked, drilling contractor hired to drill a well, and operations began after all necessary permits were obtained. The first hole, less than 5000 feet deep, had a show of oil but had to be abandoned after a heavy flow of fresh water was struck. A second hole was drilled, but it was also a dry hole yet with a strong show of oil. But the hole caved, and the consortium—all by now Steve's friends despite the failures—decided not to drill more holes. So this fourth swing at a million dollars also failed.

Steve thought to himself, "Why can I find oil and/or gas for companies who pay me but will not allow me to participate, but the moment I develop an interest, I drill dry holes?" Other consulting work for a group of Texans involved assembly of 200,000 acres in southeastern Utah, Wyoming, and Oklahoma on retainer and a two percent overriding royalty. Two dry holes were drilled whose locations were chosen by a principal investor who considered himself a geologist. One well, a cable tool hole, blew out at a rate of three million cubic feet of gas daily, and a rig fire resulted in the abandonment of the project. A fifth opportunity to make a million dollars down the drain.

Some ten years before Steve's retirement as a professor, a well-known oil man from Houston asked him to assemble data on what looked to be a major oil field in the making in and surrounding the Aneth area, later a major oil field in southeasternmost Utah. After checking out the man's credentials and receiving a letter of intent agreeing to pay Steve $10,000 for each well drilled plus a five percent overriding interest, three months of work resulted in a report showing that at least 20 well sites had all indications of success in and surrounding the Aneth area. Within two weeks of when acreage assembly was to begin, this oil man, whose father had been very successful in developing a profitable oil company, was declared insane and put into a sanitarium where he died. The acreage was later acquired by several major companies, and 18 of the well locations Steve had chosen were found productive of thousands of barrels of oil per day. The new field became part of the 960-well Aneth Field. Satisfaction doesn't put money in the bank; neither does geologic success by others. This was as close as Steve had come, but the sixth try at a million dollars was another failure.

Other lucrative contracts for exploration in the Rocky Mountains involved five major companies where Steve's income from fees totalled over $250,000. Of this he saved and invested as much as he could. The research resulted in a major success in two projects, but Steve was not allowed to participate for an interest. However, in all five cases, major publications resulted, and by such publication, plus his excellent teaching record, the university approved his elevation to a Research Professsorship. By this time, his estate totalled some $400,000 in real estate, stocks, and mutual funds, yet he was still far short of a million. In fact, Steve and his wife forgot all about his earlier ambition and spent much time helping their children acquire property and establishing scholarships in geology at their liberal arts alma maters.

In 1970, a man from Fort Worth appeared in Steve's office at his home in Boulder, introduced himself, and said, "I want to start an oil and gas exploration company to acquire federal leases throughout the Rocky Mountains. Would you be willing to become a partner with a financial friend of mine, my son-in-law, and me?" "Why not?" thought Steve. The set-up involved a sizeable retainer plus all expenses, for which Steve was to select the areas to be leased, fly the areas in his own plane to check out structure and accessibility in locales where Steve's stratigraphic and structural research pointed to potential success. By 1975 the company owned 250,000 acres of leases and became a stock company with 60 investors. Steve held four percent of the stock. The company developed productive lands in Utah, Colorado, Texas, and New Mexico. By the end of those five years, the value of Steve's interest in the company was valued at over one million dollars plus four Royalty Trusts productive from the sale of operating rights on certain productive leases.

The seventh swing, taking long chances, made Steve a millionaire after he had lost his strange ambition and two years before he retired early as Professor of Geology at the University of Colorado. Success requires many tries!

Author's Note: This plausible story is a mixed one involving the careers of several independent geologists who have become millionaires in different ways. While the names and locales are fictitious, the sequence, timing, and finances mentioned could be real. To my knowledge, a great many exploration geologists, in both mining and petroleum geology, have been or are now Professors of Geology at major universities or have retired from teaching. All of them financed their own academic research by doing consulting work rather than depending on government research contracts.

Aboard the *Morrissey*—
I am at the far right; Captain Bob
is third from right behind the box.

Part Seven

BOOZE

The Tyranny of Dinner at Six
Greetings and Toasts
A Green Bottle in a Palm Tree
The Wine Maker
Gin is Mother's Downfall
Where Larry Danced

The Tyranny of Dinner at Six

Few joys in life compare to a drink or two before dinner. Taken over an hour or so while watching the evening news, a chilled tasty cocktail with some kick to it makes even bad news almost palatable. Sipping slowly to avoid chug-a-lugging, a drink brings on a glow easily sustained by a second cocktail as one watches the biased news reporting by anchormen paid far more than they are worth. The fantastic news delivery by CNN during the Iraqi War suggested that the carefully canned insipid newscasts by most TV anchors may well become a thing of the past. Pablum, that's what it is! Evening after evening, we hear and see their ideas of what the news means rather than the news itself.

The call to dinner—which we in the country always called supper and the British oft times call tea—usually comes half an hour or more before the end of the second cocktail, a dastardly tyranny that ranks with torture on the rack. A return to simple wisdom suggests that one avoid dinner altogether and simply have a third drink. But there's danger of one's liver turning to stone. One is led to believe that all excesses, called "sins" by the ultra-conservative Christians, have their own disease. Profligate sex invites the clap, or worse, syphilis; eating too much leads to obesity and wrecked knees; sex by gays leads to AIDS, and that reminds me: why did homosexuals ever abscond with the word "gay" which means happy, joyful, gleeful? A misuse of a word! Take too much exercise and you develop excess muscle so you look like a bloated Atlas; take too little and you become soft and die young; jog too much and you wreck ankles, knees and back; take too many medications and death is inevitable. The body is fully capable of healing itself, given half a chance, except for accidents.

But I stray from the tyranny of dinner punctually at six, even worse at five-thirty. It appears that mind-bending moderation is the answer to everything that we enjoy most. An old saw goes: it's either illegal, immoral, or fattening to enjoy the illicit pleasures to excess. I knew an old souse, a true lush, who didn't die until he was 88 years old, whereas a super-Christian friend of mine kicked off at 55. This reminds me of the news reporter who went to a small town to interview its citizens because they reached great ages. The town was nicknamed "Longevityville." He drove into this town in the mountains of rural Arkansas, parked his car, and noticed three very old men sitting on a bench outside the general store. He asked one gnarled old man to what he attributed his long life. The man said he never drank whiskey, got a full night's sleep, always ate supper at six o'clock, and worked hard cutting timber.

The reporter asked, "So, how old are you?"

The answer was "88 years last week."

Then he asked the man in the middle, a grizzled old veteran who had fought in World War I, "Why do you think you've lived a long time?"

"W'all it's like this young feller. I've always worked hard, I go to church regularly, been married only once, never ate too much, never smoked, and slept in my own bed instead of chasing pussy all over town," said the old man.

"And how old are you?"

"Why, I'm 92 years old."

The reporter took careful notes and turned to the third haggard old man and asked, "How come you've aged so successfully?"

"It seems like a heck of a long life, but I drank moonshine like hell was freezing over, smoked cigars and inhaled, never worked a day in my life, laid every woman I could, never went to church, and been in jail so many times I can't count," said the third old man.

Asked the reporter, "That's very interesting? How can you possibly have lived so long doing all those things? In fact how old are you?"

The old man pulled himself up from a slouch and replied, "I'm 36 years old."

There must be some kind of a lesson in that story, but I'm certain it has absolutely nothing to do with the tyranny of dinner at six o'clock.

Sherman, Florence,
Anne, Diana, Steffen Johanson, Tim, and Stephanie
Guaymas, Mexico
1961

Greetings and Toasts

Greetings and toasts are those quickie salutations on meeting, drinking, parting, or even other occasions when a brief word or phrase is in order. They should be short, as all impromptu conversations also should be. They add to the humor of a day or night and grease the oft-times squeaky wheels of stilted conversation. A greeting in a bar, a hotel lobby or on the street may lead rather quickly to an offer of a drink plus a clever pre-drink toast. If between a man and woman, after that anything can happen—and many times happily does!

Greetings people use on meeting each other are as diverse as people themselves. The same is true of parting words. I am always surprised in Spain, Portugal and in South and Central American countries when someone greets me with "*Adios, Señor*," or in New Mexico, "*Adios, Cuate*." The latter is a word not found in classical Spanish, and it means "buddy," less formal than *amigo*. In Newfoundland, those philosophical Irish-descended sages, many of whom cannot read or write, will betray their sea-faring backgrounds by saying "Do you see the land" as a greeting.

Perhaps no people have as many informal ways to avoid the formal "How do you do" as do Americans: "Hi! How ya doin'?," "How ya goin'?," "Where ya goin'?," "Ho! What's up?" Among my more racy friends: "Are you gettin' any?," "Who're ya doin'?," "Whatcha doin'?," "What do ya know?," "Hi, doc (or mac or bud)?," "Who's around?" Among oil men the common greeting is "How deep are ya?"

Among King Island Eskimos (who now want to be called Inuits) the greeting is "*Nahru-ving*" or "*Karuk-ik-pin*" (How are you?). The Baffin Island Inuit says, "*Oxche-noi*." Greetings among the Germans and French seem less demonstrative as compared to those of the language of polyglot Americans.

As to good byes, what's more friendly than: "See ya around," "Don't work too hard," "Take care," "Until then," "Don't let 'em catch you," "Be careful," "Don't spend all your money in one place," "Keep healthy," "If you can't be good, be careful," "Don't take any wooden nickels," and a thousand other ways to shake a hand, kiss a woman, leave a friend.

There is something innately friendly about Americans. If you ask some people "How are you?," you might get what you didn't ask for: a litany of ills, feelings, bad dreams, requests for handouts. So it behooves a parting shot to be done quickly. Generally, "How are you" means hello and doesn't merit any more than a return "Hello" or "Hi." My stock answer to that question ("How are you?") is "I'm here," and the recipient will laugh—which reminds me that Tom Landry once said, "You can't think while you're laughing,"so an answer of "I'm here" cuts off all thought of a mutual exchange of "Would you like to see the scar of my operation?"

I have often thought just how inane the comments about the weather are on initial meetings. If you want to find out how seldom initial greetings, salutations and weather comments are heard, say something completely nonsensical like "I just shot my grandfather this morning"

There are many toasts in many languages that I do not know, but here are a few prior to a drink together: "*Prosit*" in German; "*Salud*" in Spanish; "Cheers" and "Here's to your good health" in English; "*Adios*" in Portuguese; "*Schlanta*" in Gaelic and modern Irish; "Here's mud in your eye" among all of us red necks; "To the Queen" is British, of course; "To the King" in Walloon in Belgium; and "*À votre santé*" in French—the French really aren't very good drinkers, except for wine.

My good friend Gordon Todd, who was raised in Honduras, thinks in Spanish, and does mathematical computation in that language, came up with this one. Holding your drink in either hand, it's called the Baja Toast: "*Arriba*," hold drinks high; "*Abajo*," hold drinks below waist; "*Al centro*," touch glasses together; "*Al dentro*," into the mouth (between the teeth). This never fails to lead to a second drink, and, if the drink is tequila (Maya, Daisy or Sunshine), after three drinks there is a mighty desire to lie down and sleep. (Works great with certain racy women who don't have sleep in mind!)

ASMU Row
Marshall Islands

A Green Bottle in a Palm Tree

Tropical islands and coral atolls have large stands of coconut palm trees, a tree of such magnificence that it supplies wood, food and drink, leaves, and shade. In 1944 during a visit to Majuro Atoll in the Marshall Islands as a Naval officer choosing geodetic points for the construction of controlled aerial mosaics, I noticed bottles with a green liquid hanging in palm trees. This was a mystery which required a solution. The answer came a few days later when one of our Naval aviators was invited by a Marshallese family to enjoy what they called palm liquor. During wartime in forward areas, we could not easily obtain alcohol in any form except beer. Through ancient tradition and recipes handed down from generation to generation, a drink was concocted that went through several stages from a mild tasty liquid to a powerful liqueur. My Naval aviator friend, "Moose" Eisele, drank several snorts and promptly keeled over, unable to fly for three days.

I have researched this liquor, and here are some answers. The drink is called *Jakauru* in the Marshall Islands and *Arrack* in the Hawaiian Islands. It is made from the fermented sap of flower stems of the coconut, *Cocos nusifers*. This unsophisticated brew is simple compared to liquor made in stills. The age-old recipe produces "palm toddy", as it is called in Samoa. The palm tree is climbed by "toddy tappers," agile boys who climb 30 to 50 feet to just beneath the tip of a tall coconut palm. Coconut buds are selected, cut into with a sharp knife, and bound tightly with palm fronds. Glass jars or bottles are hung below the cut buds, and the toddy juice drips slowly into the containers over a period of several days or weeks. The liquid, when fresh, is sweet, non-alcoholic and looks like barley water but has a natural fizzy tang. Allowed to ferment to the next stage, it becomes a greenish liquid which is mildly intoxicating. This induces raucous laughter and revelry, leaving the drinker with a mild hangover. So far as I can find, the Samoans drink it all at that stage.

When I revisited Majuro Atoll in the Marshall Islands in 1981 to continue research on its geology, this coconut juice-to-alcohol sequence was also explained to me by Oscar de Brum, Chief Secretary (vice president) of the Marshall Islands Republic. It's made this way: a slit is cut into the heart of palm at the top of the coconut trunk; the juice, called *Jakaro* or *Jakauru*, drains into a bottle hung high in the tree; if left to ferment naturally, it becomes a beer-like liquid called *Jemanim*. When this is distilled, it becomes a fiery drink called *Tuba*, very nearly straight alcohol. That's the drink that laid low my Naval aviator friend, "Moose," in 1944. He learned the hard way that a green bottle in a palm tree can lead to disaster!

The Wine Maker

The Volsted Act was a tragedy only for a few months for my father. Ever resourceful, he beat that heinous attempt by Congress to regulate the morals of a whole nation. He researched making wine as soon as it became unlawful to make and sell anything alcoholic except that travesty of all travesties: near beer, a low percent concoction. It took gallons of it to get a buzz on and led many jovial drinkers to learn the art of making home brew. I sampled home brew when I was 12 years old in the cool, humid cellar of a man for whom we had roofed a shed in Winesburg, Ohio. But long before that, when I was four years old, I drank several glasses of my father's claret, put out within easy reach on a cold day while the men were butchering two hogs in our back yard. The result of that was a binge, my first, followed by a monstrous headache.

My father bought a new hogshead, a giant barrel made of maple staves with a removable lid, and placed it in the coolest corner of our cellar beneath the house. Before setting it on bricks on the earthen floor of the cellar, he charred the inside of the barrel lightly by burning fired oak branches under the upside down barrel. While he sliced whole oranges—hard to come by in those early days after 1919—my mother picked baskets of plump blue concord grapes from the arbor north of our kitchen and east of our summer house. I was detailed to pick dandelion blossoms in our backyard. They grew only in the backyard out of sight, as my mother would not tolerate them in the front yard by the road for all passers-by to see. Into the barrel went the whole mess.

I never did know the proportions, but into that right-side-up barrel went orange slices, grapes and dandelions with some yeast and lots of granulated sugar plus water from our 20 foot deep open well in our front yard—beautiful, sweet, tasty drinking water. Being only four years old, I expected wine right away. Disappointment reigned as my father would go into the cellar each evening after coming from the warehouse, which was then on the brick-paved South Mad Anthony Street in Millersburg. He stirred the contents of the barrel every day for several weeks, dipping out somewhat more than a taste each day to check the progress of the wine.

After what seemed to a four-year-old boy an interminably long time—several weeks—I watched him take a fine cheese cloth, load it with the mixture in the barrel, and hold it over a long funnel stuck in the bung of a small barrel lying on its side and blocked on a table so it wouldn't roll. The odor was heavenly as the bluish red liquid ran into the small barrel, which had a spigot on one end. "Ah," thought I, "now we get a drink of this good stuff." Of course I couldn't think of such a word as ambrosia because I couldn't speak English—only "*Ach es schmoked gut.*" But it was not to be. He drove the stopper into the bung hole and said in German, "Now we let it sit for several months." But he did let me have a taste, and my mother insisted on a single glass full. No doubt my father had more as time went on; his excuse was that he had to see how it was doing. So it was another long wait, and I forgot about the wine as the summer wore on.

One evening in the fall, my father went down into the cellar with a cut-glass pitcher given to my mother by her own mother. Up he came into the dining room off the kitchen with the pitcher partly full of the loveliest ruby red liquid I could ever imagine. My mother had a small glass and so did my 12-year-old brother. My father, with a flourish, poured himself a whole tumbler full, and I had a very small tin cup full. Before supper we sat around the dining room table which my parents had bought at Vogt's Store for $4.00 when they were married in 1906. We smacked our lips as we tasted that lovely, sweet wine. My mother loved it but rationed it for herself to one small glass on weekends, allowing me only a taste from her glass. I'd guess, now 70 plus years later, that the small barrel held ten gallons of wine, superbly aged. The first turn of the spigot drew off a bit of sediment which went right over into the hogshead in the spring of 1920—already charged with the makings for wine to be completed by that autumn.

Until the Volsted Act was mercifully repealed in 1933— "The only really good thing Franklin Roosevelt ever did," said my father—we had that beautiful ruby red claret wine in our home. The Volsted Act was thwarted! Long live those who take a good long drink of homemade wine.

My brother Owen with Dad and Mom
c. 1910

Sailing in the Virgin Islands with friends—
We are on the right.
1968

Gin is Mother's Downfall

Dean Martin once said, "Of all the remedies that won't cure a cold, booze is the best." The British favor plunk, a cheap wine served hot, to make a bad cold bearable. Mexicans dote on tequila or gin peppers, which are jalapeños (hot peppers) added to the bottom of a bottle of gin for six weeks and well shaken up. Miserable as a cold may be, it is always a good excuse to become rubber-legged and fall asleep. But when awake, George Jean Nathan avowed, "I drink to make other people interesting." Jim Cook, a sailor from Florida, swears by a drink called Anchor Down which he concocts for happy hour after a long sail. Here's what it is: One and a half ounces rum (the best, not the cheap white stuff), ½ tsp. sugar, ½ lime (or ½ tsp frozen lime or 1 oz. lime juice) plus water and crushed ice. That drink wakes up the cockles of the heart and makes all women look sexy! But my brother, Wilmer, swears by 1 ½ oz. gin with half-and-half tonic water and cranberry juice. He says that's a cure for heart disease of all types.

The Hawaiians long ago invented the Mai Tai, and here's the recipe for two people: 2 oz. white rum, 2 oz. dark rum, 2 oz. orange liqueur (triple sec), 6 to 12 oz. pineapple juice with a float of 1 oz. of 150 proof rum on top of crushed ice. Add a pineapple slice, a maraschino cherry, a piece of lime, and a sprig of mint. If you can get through all the vegetation and fruit before your thirst disappears or you get tired of chewing, it's a great drink, especially at Trader Vic's. But beware of two of them. They'll make you falling down drunk.

A friend of mine gave me a recipe for what he called "Cherry Bounce." Take 1 quart of cheap(!) bourbon or Canadian whiskey, 1 quart of sour red pie cherries with juice, and 2 cups of brown or unrefined sugar. Shake them up together and put in quart glass jars sealed tight. Let it set for six weeks, strain it, bottle it, and serve it only in a glass over ice. It's best to loosen the cap of the original container each week or it may explode. The result is a drink very much like Danish Cherry Herring!

At the Hotel Maya Bar in Antigua, Guatemala, I learned about one exotic native drink called the Hunapucocktail: ½ banana, 1 oz. lemon juice, 1 oz. orange juice, 1 oz. pineapple juice, 1 oz. rum, and 1 oz. orange liqueur; blend with ice. That drink should satisfy the need for vitamin C for a whole day!

One can expect almost anything from people in central Ohio. My brother-in-law told me about a drink called the Fuzzy Navel. No amounts are given because after you drink one you don't remember. It's made of peach schnapps, vodka, orange juice, and ice thrown together in a blender. The recipe is up to your own imagination. One of his friends uses gin, another makes it with bourbon!

In Australia the Abos make a beer called "yacona" out of kava root. It's a non-alcoholic drink that numbs the lips, tongue and throat. It also immobilizes a man below the waist! Great for killing a sore throat! Australians must be the greatest of all beer drinkers, for they have different names for beer containers: a Stubby is an 8 oz. can of beer, a 7 oz. is a glass, a 10 oz. is a Middy. Too much beer and the Aussies get riled up rather quickly. In the Whitsunday Islands on the central Barrier Reef, the favorite local rum is Red Spot which is drunk with prune juice. Belgium is the only country I know of that has at least 327 kinds of beer made by myriads of breweries!

Any good doctor will tell you that having alcoholic beverages after the age of 40 is not only good for the body but might be good for the soul as well. Doctors warn, however, not to drink every day. Drop one day per week. Alcohol is a drug, a social drug easily burned up by the body—unless one drinks too much too fast. The alcohol of favor is vodka or gin, as these two have no congeners,which give color and taste to bourbon, whiskey and rum. Congeners are unhealthy and, like those in red wine, cause the synapses and nerve endings to pull apart; also red wine causes petrification of the liver if taken in excess. The truly social drinks are beer and wine, with white wine the lesser evil as compared to red wine. Where beer ever got the reputation of being the drink of low brows is difficult to determine. The ancient Egyptians, the high brows of pre-Christian days, brewed beer and made wine from grapes, as well as alcohol from grains, thanks to the discovery of yeast. As to wine, the idea that it's the drink of the elite is fostered by the French, whereas the Californians, the Spanish, and the Italians drink wine for fun. The hardy English have their stout that they brew almost to the exclusion of Pilsner and lager which they import from Holland, Germany, Denmark, and Belgium. All in all, alcoholic drinks lead to many frolics in many places, to happy hour with two drinks for the price of one, to conviviality at the end of a hard day's work, to rousing raucous cheering before the TV and its sports events. As a friend of ours in London declaimed, because his wife is hooked on gin, "Gin is Mother's downfall." But it's an old desire, for man has the same affinity for gin as his tetrapod ancestors had for juniper berries in prehistoric times.

So how do you tell if an educated man is drunk? He might say, wobblingly refusing one more drink, "I have had an elegant sufficiency. More would be a superfluous abundancy." The humor of the classical lush is at times overcome by pugnacity beyond belief. Fortunately, until the liver turns to stone and as long as there is moderate drinking, humans will find themselves and others comical, as well as brilliant.

Sherman A. Wengerd—
"The Great Snake Tamer"
Honaker Trail, Utah
Summer 1951

Where Larry Danced

It's an unassuming little place, tight by the highway between Cortez, Colorado, and Monticello, Utah. It's called The Frontier Bar, and it is on the southwest side of the main street in the village of Dove Creek, Colorado. This was the scene of many stops as we shook off the dust from the red desert lying to the southwest. Dove Creek rests on the Blanding Plateau and grandly proclaims itself to be the "Pinto Bean Capital of the World." The rolling fields of red soils, blown in the wind from the Southwest deserts and added to the pyrite-turned-to-limonite-stained soils of the Dakota coal seams, create a rich sandy soil covering thousands of square miles of southwestern Colorado. Rainfall here is 20 to 25 inchesper year, and some of the flatter acres are irrigated by giant sprinklers. Bean elevators dot this high plateau countryside all the way from Cortez to Monticello. Giant machines till, harrow and furrow the thick, rich, red soil, and in early autumn large harvesters reap this most elegant and tasty of all beans: the pinto.

But beans are not what this is all about, except to note that although some of the people are non-Mormons, most of these bean farmers are know as Jack-Mormons, hard-working, hard-drinking rather unorthodox Mormons. The Frontier Bar is one of their favorite watering holes after a week of hard work in the bean fields.

On an early Friday morning one fall in 1948 during the bean harvest, an unusual assemblage of men left Durango, Colorado, in a Cadillac with Kansas license plates. Leon Derby, President of Derby Oil Company; Larry Stubbs, landman for Sunray Oil Company; Bob Painter, Sunray geologist; and Sherman Wengerd, consultant to Sunray in the Four Corners Region, were bound for the Mexican Hat Oil

Field 65 miles southwest of Monticello. Leon had an option to buy the entire oil field. The trip was to convince Sunray officials that they should join Derby in revitalizing this old field. Discovered in 1908 and one of the first oil fields in Utah, Mexican Hat had the distinction of being not only the smallest field in Utah but, by 1948, the least productive. Located on the flank of a downfold rather than on top of an upfold, Mexican Hat was at best peculiar in that the paraffin-based oil was of high gravity and full of gasoline, thus very valuable. At the field the quartet turned on an abandoned tank and collected some of the oil, noted one well being drilled with a cable tool rig, and saw that the driller was fast asleep in the doghouse.

The scenery in this part of Utah is among the most spectacular in America. As they hiked through the sage plains locating old wells, noting the rocks and taking color photos, Leon all the while was declaiming how Sunray-Derby could make a mint of money by redrilling the field. They could utilize the new technology of acid-fracturing to boost the production and haul the oil to refineries in Grand Junction or Salt Lake City in long tanker trucks and watch the money roll in.

About seven in the evening, as they drove back through Dove Creek on the way to a late dinner in Durango, Larry spied The Frontier Bar. "Stop!" said he. "I'll treat." Dusty, thirsty and hungry, the four tramped into the bar by the side of the road. The initial idea was to have a beer and drive on, but all of them noted the pretty red head tending the bar, whose husband was somewhere out west finishing up the harvest in a bean field. One round led to another, a radio played music from a station in Salt Lake 300 miles to the northwest, and Larry decided to dance. Inspired by acclaim, he whirled, crouched, leapt up in graceful pirouettes, ran around in circles with his hat pushed back on his head, alternately laughing and singing at the top of his voice. It is entirely possible, indeed probable, that all four joined in this bacchanalian performance, cheered on by the redhead who kept plying them with free drinks. It is difficult to remember!

To this day, The Frontier Bar is passed hundreds of times by geologists and engineers with the comment "That's where Larry danced." The state of Colorado should put up a tourist plaque to commemorate the occasion when Larry tore up the floor in a dance the likes of which was never seen in Colorado before—and may never be seen again.

On our honeymoon

Part Eight

MARRIAGE AND THE OLDEST PASTIME

Option and Choice
One Marriage
Jowls and Withers
Long Haul to Freedom
Sex is for Everybody
Shirt Tails and Women
A Slice Off a Cut Loaf

Option and Choice

There is a vast difference between an option and a choice. Options are the possibilities whereas choice is an active decision among many options. The main leverage one has is to arrange their life in such a way as to have numerous options on everything one wants to do. Take the example of choice of a man as a husband or a woman as a wife. One has options involving the good, the bad, and the pretty and the not-so-pretty, the big or the little, the smart or the stupid, the rich or the poor, the fat or the skinny, the tall or the short (after all, big, fat and tall may not always go together!), the laughing or the dour, and so on. But it is the general moral thing to do: choose one mate and stay with that choice, as marriage is the one contract that is supposed to be for life. This means the choice among many options should be the right one, even though annulment and divorce are common ways to abrogate a poor choice.

Once married, options of certain types are no longer possible, but many other choices must be made among options agreed upon by two people. When children come along, themselves options, the situation becomes complex indeed. First what brand of child do you want, even though you take what you get. I'm told that wise Jews have worked out a sequence of events that virtually guarantees what the sex of a child will be. Which schools for the children? Until lately which primary and secondary school was dictated by where one lived. The exception: private schools over public schools. One must sort over many options if finances permit.

There are numerous options involving one's career and where one wishes to live and work. My father told me to prepare myself for many different kinds of work, several types of careers. Keep your options open, for who knows when you'll run into a boss who is an ass and completely impossible, a job or position you don't like, or a place you may be sent where you absolutely do not want to live. I know over one hundred acquaintances and friends so well educated that they could pick and choose careers. They were so intellectually agile and mentally adaptable that they could be, and were, successful in several careers in sequence, and never missed a beat or a pay check! Options are the wellspring of successful men and women.

It is doubtful if there exists anywhere in the world an environment so conducive to numerous options as the United States of America. Any political-economic system under which a great number of incapable and lesser capable people can be successful in life (make a living, raise children, stay out of debt, and be happy) must be what the Lord meant when he said, "The meek shall inherit the earth." And where does this interesting condition exist but in the USA? All freedom is costly in time and money, so our multitudes of options have been difficult to gain and maintain against bureaucratic restrictions, whether by government, big business, bankers, large universities, or petty martinets in all fields of endeavor. Generally speaking, man's inhumanity to man is probably nowhere better developed than in the large cities. The crowded rat syndrome is real. To become a cog in a gigantic machine limits options and deadens the imagination. If options are limited by precedent, eliminate the precedent. Any mentor who insists upon precedence to maintain a position cannot long maintain an untenable position in view of changing conditions. It behooves us to develop our own options and to act.

Arnold Toynbee said that "An active minority can control an inactive majority." It soon becomes obvious to an alert person that those who create their own options are a minority of one. An option toward action which the majority doesn't understand can be dangerous. To get options on certain actions may require one to join a majority. One either leads, follows or is asked to get out of the way. It is good that life is like that, or we would all be on the same low level of mediocrity, a stultifying condition. Sir William Gilbert of Gilbert and Sullivan said, "When everyone is somebody, then nobody is anybody"—a truism which means simply that options to be different may no longer exist. An option may be chosen that leads to total security. Prison is the only place where complete security exists and options are absent. How dull!

June 12, 1940

One Marriage

Every successful marriage starts out with a number of agreements mutually derived. An old saw comes to mind: "I chased her until she caught me." We met in April of the year I earned my master's degree in geology from Harvard and she received her graduate degree in library science from Columbia, both in June. However, we didn't start dating for several months.

While I was in debt to a bank at home to finance the first year of graduate work, she had a library job and was saving money. While courting her, we discussed many things. As our acquaintanceship progressed, I found she was right-handed, whereas I am left-handed. A dextral wife with a sinistral husband—so far so good. She was farsighted. I thought that a good omen and considered that to be why she put up with me. I was slightly nearsighted and didn't know it, having passed U.S. Navy examinations for flight training two years before I had met her. We discussed sex rather casually, finding that we both wanted children—preferably four—both recognizing, without undue experimentation, that we were sexually suitable for each other. We discussed money, and she let me know somewhere during our dating game that she wanted a husband who handled the finances, would allot her an allowance which she could spend any way she pleased, and would do the investing with her concurrence on any large investments should we ever be so fortunate as to have enough money to so do.

My wife came from a religious and academic family; I came from a business and contractor family with a religious mother who wanted me to go to divinity school and a father who insisted that I become a medical doctor. (The clashes with my father over career and six other major decisions in my life are told elsewhere.) We found that the finances of both our families were nearly identical, which had no bearing on our relationship; hence, it wasn't a rich girl marrying a poor boy or vice versa. We were married in 1940, and I had to borrow $500 from Dad just to finance a honeymoon trip from Massachusetts to Oklahoma to begin work as a geologic trainee with Shell Oil Company.

We found we almost automatically made a number of long-term decisions about many small matters. Double beds or twin beds: no hassle. But, when twin beds, she took the left-hand bed facing the head of the twin beds, or the left side of a double bed. I leave to you the reason for that choice! A car with a wide front seat or two bucket seats? She married a young man with a second-hand Hudson with a wide front seat—no big deal. Who pays for what? I take care of all car and travel expenses and taxes, plus housing; she takes care of all our food and household costs and her clothing. I pay for and choose my own clothing, although she is not bashful about stating her opinions, as I am not bashful about recommending what we have to eat. I found that her mother didn't really like to cook and do housework, quite different from my mother, though both had maids. My wife could cook steaks and scones—that's all—so I had to teach her many Pennsylvania Dutch cooking skills, such as how to fry an egg. All this led to many hilarious differences of opinion as she learned how to cook. A dusty house? In Oklahoma, Kansas and Texas, who cares about good clean dust? There are things much more important about a good wife than a super clean house and gourmet cooking.

Did she like to camp, sail, take voyages, go on long trips, live on a low allowance for some years until children came along? The answer was yes, so I knew I had a great person to be the wife of a geologist. Of course, I had a leg up, involving a person able to live with a geologist, as her father was a geologist. One slightly detrimental matter struck me: she had no brothers, so I had to teach her about brothers.

We got to know each other and became staunch friends through the thick and thin of nine moves while with Shell for two years before joining the U.S. Navy. Several weeks after we were married, my boss at Shell Oil Company asked us to live in his house while he and his wife took a three-week vacation. It was a palatial mansion in Tulsa, rather than the one room second floor garage apartment, which we then occupied, with a bathroom in one corner, a kitchen in another corner, and a bed in the middle. You bet! We jumped at the chance. One Sunday as we were sun bathing on the hidden driveway of my boss' house, we planned our family and chose names for all four children. There were to be two boys and two girls. Something went awry by 1951, and our last child, who was to be a boy, was a girl, so instead of naming the requested "him" Stephen, we named the arrived "her" Stephanie. Because the male determines a child's sex, I couldn't blame anyone but myself.

So on and on, through more than 50 years of marriage we were considerate and developed deep respect for each other. Of course there were arguments, but we never both got angry at the same time, and the sun never rose on an unsettled difference of opinion. A conclusion is in order: marriages may be made in heaven, but to keep one in good repair takes a lot of maintenance work on earth. Stephan Zweig had it right when he said, “Love needs presence if it is to endure.”

Married 50 Years

Jowls and Withers

This title may throw you off track, for it's about successful marriage. How to stay happily married to one woman has intrigued many men, whereas others don't care, thinking a change highly beneficial now and again. The marriage contract is really the only one for life and is easily broken. There is one friend of mine who simply doesn't bother with marriage. He's had six or seven different women, all great looking, without benefit of or detriment to that piece of paper. It's intriguing to contemplate how he does it. They live with him for several years. They do his clothes, cooking, travel with him, and suddenly we find he has a new one. Well, not exactly new, but different!

This idea that one could remain married to the same person for fifty years—and why—came about in church one day when the minister, in front of God and everybody, asked a couple to stand up and tell just how they stayed married for 55 years. Of course it was none of his darn business, but the wife simply said, "Work at it"; the husband just hunched his shoulders and said nothing. This set me to thinking about how I would have answered his question.

It occurs to me that many marriages break up because one or both of those in supposed wedded bliss snore! Yes, snore! I believe it's a bad habit, a concept a doctor friend of mine says is not true. It has, he says, something to do with labials— whatever they are—in the nose and throat.

Sexual compatibility, any sexologist will tell you, only requires generally good health, a willing partner, a good imagination, and a strong prostate gland. As for the long life necessary to enjoy sex into old age, it's clear that a strong elimination system is very important; also a strong imagination, a desire to please, a powerful lymphatic system, a strong heart which drives an efficient circulatory system, and a bit of voyeurism on the part of the man with a touch of narcissism by the woman. It should not strike you as strange that I should mention sex as the prominent factor in a successful marriage. Perhaps it may occur to you that my wife and I may know something about this—and why.

But there are many other reasons long marriages are possible: mutual respect, kindness toward each other, mental alacrity, many diverse interests, and separation for some hours each day, such as the husband in his shop, dark room, office or study and the wife in her music room, sewing and weaving room, kitchen, reading room. One needn't have a gigantic home, only one in which rooms have multiple uses. If both enjoy music together, traveling together, eating together, sleeping together (remember, NO snoring!), sex, and having children together, what can possibly cut such a marriage asunder? All husbands worth their salt like an earthy woman, and women seem to like direct action doers for husbands. Lastly, there can be no totally successful marriages in a state of poverty. Both must agree on financial matters: mainly how to save, spend and invest money toward the day of retirement freedom—as well as during retirement!

Long Haul to Freedom

Once the decision is made to have children, a couple begins to lose certain freedoms so much enjoyed in early years before the first child arrives on the scene. Before birth control was generally available, such freedom was negated in part by ever-impending pregnancy. A baby, eagerly sought, doubles one's responsibilities, and costs begin to mount. It's good that children arrive when one is young, for the joy makes up for any loss of freedom. One is generally too busy to notice, as the work-a-day world and the beckoning future fill time so completely while children grow to adulthood. One couple I know said they thought one should be able to freeze children between the ages of two and eighteen, the most difficult years.

Four children in our family came between 1942 and 1951, and it wasn't until 1972 that we finally could spend our monthly pay check on ourselves and our own desires. Those 30 years are now like a dream. They passed so fast that we don't know where they went until we look at our grown children, all married, all through college, starting families of their own. Now there are eight grandchildren of all sizes and even a great-grandson, an exact duplication of his father at the same age 26 years ago. Perhaps that's the only immortality we will ever know. If so, it's enough!

But financial freedom, grand as it is, can never match the joy of watching children make their own way: building homes, travelling, camping, having parties, having their own children, working hard, saving and investing for the future, and planning their retirements. Of course, trying to be an example and guiding them with advice that they seemingly didn't take was tough. I am amazed at times to hear them counsel their own children with the exact words of advice that their mother and I gave them. They did learn—and we thought they weren't listening!

They had fiery independence as they were growing up, and they made many of the same mistakes we made. This shows the essential cyclicity of the generations as they pass through what many people, fatalists that they are, consider life to be: "This vale of tears."

One is reminded of the story about the 18-year-old boy who was convinced his father was abjectly stupid until the day the boy became a man of 21, and he realized how smart his father really was. Wondering to himself, the young man said, "How can my father have learned so much in such a short time?"

Freedom isn't really lost with the advent of children. We only gain different responsibilities. When parents see their children meeting responsibility with éclat, that's a freedom beyond believing.

The family
June 1990

Sex is for Everybody

There are just too many people in the world to have equal opportunity for all in all fields of endeavor, but as long as the ratio of men and women stays close to 50/50, every woman should be able to have her own man, or, as Yogi Berra would aver, vice versa. That some men and some women are greedier than others and that there are many in both sexes who care not a whit for the opposite sex cannot be doubted. The latter may be true for asexuals as well as homosexuals. That the former is true may be laid to the fact that there are great variations in monogamous magnaminity—or, put another way, some are cursed, or blessed (some would say), with overly powerful sexual drives. It seems that humans are the only animals who enjoy sex purely for fun.

The overtly religious are non-believers in the truism that heterosex is actually no one's business except the participants. Havelock Ellis thought that marriage was really only a legalization of prostitution. If so, of what value is a piece of paper between two consenting adults of the opposite sex? These statements may shock many strait-laced people who aren't prudes at all but are the descendents of generations of puritanical morality. Of course that jokester Jacques Offenbach had a different view of woman: "All women are a nuisance. But at least with one's own one can always go for a walk."

Two of my very old acquaintances, now long dead, used to quote two old saws about rape, which is forced sex, an expression of hate, antipathy and the warped criminal mind: "When rape is inevitable, relax and enjoy it" and "A woman can run faster with her dress up than a man can with his pants down." These concepts of sex are crude warps of an activity that is the wellspring of all human endeavor wherein two people enjoy each other in their own versions of love. Love is a mental conjugal arrangement whereas sex is the physical expression. Lord Byron in "Don Juan" believed, "Man's love is of man's life a thing apart; 'tis woman's whole existence." How's that for male chauvinism?

There isn't a sexually active man alive who hasn't had what some would call lurid dreams about sex with someone other than his own marriage partner. It happens to women more often than they will admit. A dalliance with a lovely woman, a complete stranger, one whom I had never met, was a dream of such intensity that I remembered her name to be Mrs. Curtis Pease Jackson. Note her husband's name. In the dream I never learned her given name, and I truly hope no such name exists. The very strangest thing about this is that she looked exactly like my wife! I dare psychologists to analyze that one. Ben Franklin, that august purveyor of advice to young men wanting to marry, once said, "Marry a woman who is an angel on the street and a devil in bed"—eminently superior advice, for that is exactly what a prostitute is not. Fairfax had it right when he believed a broad bottom and a warm heart to be better than a kingdom.

A part of the mental freight a man carries is that of the attacking dame who throws herself at him. This is the glitter of the glands, an illusion that makes thinking about sex so appealing, especially if it involves a demure, apparently shy woman who seethes unnoticeably. The Marine motto, "semper paratus," is a valid piece of advice when confronted with such opportunity. However, a woman scorned is dangerous; the phrase is "Heaven knows no fury like a woman scorned."

Lord Hamilton in the 16th century said of the sex act that "It is a ridiculous position of momentary pleasure and abominably expensive." I would venture that Lord Hamilton was an unfeeling man of the slam-bang school who didn't really wear the pants at his home. Cato in 215 B.C. went a different direction when he stated, "Suffer women once to arrive at an equality with you, and they will from that moment become your superiors." This is the approach of a male unsure of his ability both sexually and personally. The French, always appreciative of women, know "Women are different from men. *Viva la difference*!" Men like Balzac and Falstaff recognized the beauty in women as objects to revere for their physical attributes. The more laconic will say, "Women: you can't get along with them and you can't get along without them."

Socrates, that free thinker killed by hemlock forced on him by his less philosophical peers, "shed his seed upon the ground" in defiance of his bitchy wife, Xantippe. Women who withhold sex as a tool to get what they want may or may not be worse than women who use sex for money. The Bible is filled with sexual innuendo and the beauties of sexual love, especially in the Book of Solomon as well as in some of the Psalms. Why then do so many religious women believe that sex should be endured for procreation only? Formerly it was fear of progeny, later fear of disease. Always it may have been teachings that to have sex was dirty, immoral or an act of the Devil. Stefan Zweig, that wise author who quoted Balzac, that agile lover of many women, mentioned this: "Passionate love in a woman is characterized by a boundless capacity for surrender." What a lovely thought!

Senator Alan Simpson of Wyoming, in a TV broadcast of an important committee hearing, once pulled this western quote about a wanton woman, "She's a red-headed, double-breasted mattress thrasher." The reason that sexual activity has such a rich vocabulary among the couth, the uncouth and everybody else is that it can be the most joyful act of the entire human experience if shared in complete mutuality. Of course there are always those spoil sports like Lord Chesterfield—possibly misquoting Lord Hamilton—whose concept of sex was "The price is prohibitive, the pleasure transitory and the position ridiculous." So whom do you believe? The answer is yourself! Have at it and have fun with a willing partner! And don't forget the physiological advice: "If you don't use it, you'll lose it!"

50th wedding anniversary celebration

Shirt Tails and Women

I wish somebody would tell me why women, especially young nubile females, at times like to wear nothing else but a man's shirt? There is nothing quite as silly looking as a man wearing just a shirt with tails flying. But a shapely lass wearing just a man's long-tailed shirt is a sight for roving eyes—exactly the effect a fetching female seeks to elicit. Nothing is more inviting! Of course it's rather gauche to even question why, when every male knows that shirts button down the front—and as easily unbutton. I've asked my daughters those questions, and right there I experience looks which tell me that they think I'm at least three generations behind the times.

First a man's shirt is big, roomy, flops nicely when a woman walks along; the tails can be flipped flippantly away from searching males. Enough shows to egg a man on, and the shirt covers vital parts not popularly shown on beaches and by swimming pools—except at Club Meds and on nudist beaches. In fact, this much body, especially of males, is not a pretty sight as compared to what one can't see covered by a man's shirt on a shapely woman. Bob Hope once said, "If you pass a girl and she looks around, follow her." It doesn't take long for a young virile male to learn that a girl wearing a man's shirt is becoming and perhaps barely available.

Intriguing as well is a young female wearing a man's shirt and a man's snap-brim felt hat: the kind Indiana Jones wears in the movies, the kind of hat a man wouldn't have been caught dead in just a few years ago because it was so out of style.

Television ads titillate the imagination when a comely blond wears such a hat and a man's shirt, unbuttoned, with a come-on look on her face, advertising razor blades! With glittering eyes, tongue showing between pearly teeth, head tilted forward, eyes turned up just a bit and wide open—what are men supposed to think about? Razor blades? Or perfume?

Chief Justice Oliver Wendell Holmes, who, at the age of 80, passed by a beautiful woman on the street on a summer's day, mused to fellow barrister Felix Frankfurter, "Ah, to be 70 again!" And that rogue, the British author and playwright George Bernard Shaw said, "Ah, youth. Too bad it's wasted on the young."

Lives there a man so dead that he does not dream of past conquests when he sees a pretty woman? Remember what Ben Franklin observed about women—and that man knew what he was talking about—"No woman is old below the waist." His advice to a young man by letter concerning marriage is a classic. To old Ben there were no old or ugly women in his coterie of conquests around and in Philadelphia. I'll venture he learned a lot about sex when he was our ambassador to France during several critical years of our American Revolution.

As for sexual attractions and conquests when one is older, do you suppose the comment by Larry Anderson would apply?: "You're young only once, but you can be innovative forever." I wonder how a shapely older woman would look wearing just a man's shirt and nothing else? Worth a try!

To those resentful of lost powers involving women, such as some of my friends profess sadly, it's best to remember what Buddy Hackett once said: "Why should I be sulking about my resentment when the other guy is out dancing." One rather garrulous friend of mine still has the spirit to say "All I can do now is look, but I sure enjoy what I see." His wife would clobber me if I told you his name. Maybe he should have his wife wear just a man's shirt some cozy evening. After all, they have children, and there's no doubt that the intelligent young man becomes father to the creative old man!

Dancing at the wedding of
granddaughter No. 1
November 20, 1993

New Caledonia
1991

A Slice Off a Cut Loaf

Any woman who is picked up in a bar can expect to have happen to her what she thinks will happen to her if she is picked up in a bar. Many people are quick to give advice on sex, but one piece of advice to remember is never to get your tail where you earn your groceries. When charity is discussed and a pitch is made for a contribution, some women will say "we gave at the office," making one wonder what they gave. It may then be unnecessary to mention that it's logical to be blamed for sexual harassment if you try to seduce someone who works for you or with you. One lady I know says she thinks many women attract rapists because of the way they dress. Many women's demeanors have led to passes sought whether for fun or pay. Art Mosher tells of a rakish friend of his—a man some sixty years old—who performs an experiment every once in awhile just to see if his technique has lost its effectiveness. He walks up to a pretty woman on the street and asks, "Would you like to go to bed with me?" Now, Art's friend is sort of a handsome devil-may-care type with a roving eye who was a pilot in the Air Force, and no doubt he's careful to pick a woman who he thinks may like a bit of fun for an afternoon. Art asked him, "How does that work?," and to his amazement his friend told him, "You'd be amazed how many of those I ask say 'yes'!" Zorba was right: "To be alive a man must undo his belt and take a chance."

Bud Dessart, a petroleum engineer who has worked all over the world and has done lots of exploration between the sheets, once told me, "Whores, taxi cab drivers, old women, bartenders and little children are really the only sympathetic people in this cold world." But people need to limit amatory athletics to the non-professionals. He said they were more fun, and, further, he believed the old cliche works: "Candy is dandy but liquor is quicker." Bud would wake up after a night at a night club in the black section of Luanda, Angola—among other places—and say, "Today I'm bleary eyed and bushed from tail." However, there are older men of low libido who say, "A good crap is better than a bad lay."

Every man with imagination, happily married to a willing partner who has moderately good health, knows that his wife is, in her psyche, a bit of the whore. Look at her joy when he gives her a special allowance in green backs or an expensive gift.

There is no doubt that marriage is a sacred institution, and many men and women are locked up in it as in an institution. But fiat and custom dictate that one should have only one wife at a time. The French solve that dilemma by having mistresses just as kings had concubines. Some movie stars partake of many men, and some men many women, whether married or not. Freedom to have many marriages is easy. Note Johnny Carson, Mickey Rooney, Zsa Zsa Gabor, and Liz Taylor, although it's known to be very costly. An old friend of mine many years ago lost his wife and took up with a gold-digging bitch. His sons said, "Listen, Pop, we'll pay for all the women you can handle. Don't marry her!" Like many fathers, he didn't listen to his sons, married the woman, and she cleaned him out, tossed him out of his own home, broke him, and he died of a heart attack all within six months. This leads me to quote a Harvard buddy of mine. When asked if he would remain faithful to his wife after he married her, his retort was "A slice off a cut loaf is never missed—and there are lots of loaves available." He continued: "Street cars come along every fifteen minutes." Finally, it may be said of many a woman that her legs, formerly parallels of virtue, are now pleasure bent and wellsprung; also her shoes have round heels.

Flight over Haleakala—
Great scenery

Part Nine

KEEPING WELL

Reveries and Dreams
Blood Pressure
So You've Quit Smoking
The Old Diseases
The Long Way Back
Old Ladies
Never Run Downhill

Reveries and Dreams

In the scheme of things and ideas involving the human mind, nothing so much intrigues people as the flow of thought while relaxing. To allow this flow to operate—in fact, to encourage it—takes no real concentration. If one records sequences of thoughts, the very act of recording stimulates an amazing range of subjects. As one thought engenders another, logic becomes twisted to illogic, the heart beat may quicken, the temperature may rise, and muscles may tighten to create panic. Conscious thought must inevitably take over, and such dangerous reveries can be stopped by getting a drink of water or something stronger.

The poet, prophet and prose writer may use reveries as a tool. If alcohol is used, reveries may turn into visions of snakes, elephants, crocodiles. Edgar Allen Poe wrote his most penetrating poetry in an alcoholic haze. For others, alcohol leads to violent reveries, at times excessive enough to be dangerous even to friends. Such reverie-driven activity may lead to efforts at levitation, dream-like hallucination and self-destruction.

It has been said that dreams are the subconscious musings of some inner soul. Dreams in color, rare as they are, are most entertaining. A psychologist once told me that people who regularly dream in color border on insanity. Physiologists will tell you that high blood pressure causes dreams of normal-sized things and people being seen as miniatures. Recurrent dreams are normal and most are anxiety driven. Reveries are a kind of conscious dreaming and, in my own cases, are most often humorous. I have many times had humorous dreams in which my laughter has awakened me—and also my wife, to her consternation! What does that mean? Psychoanalysts will use long incomprehensible words to classify most thoughts, reveries and dreams. I am convinced that they really know little about the human mind because they seem to forget that all thinking individuals are vastly different. So are the unthinking ones!

The human mind can be guided in a very meaningful reverie, especially if the mind resides in a healthy body. If one can measure another person by the size of what irritates him, it may be possible to analyze one's own dreams (unconscious) and reveries (conscious) and to harness a power that will guide one in life. A professor of mine in graduate school, Dr. Reginald Daly, once told me that he tries to solve difficult riddles of the earth's origin by gathering as many facts as he can—even divergent concepts—and then getting a long night's sleep. In the morning, he has answers. He then tests and retests until he is satisfied that his mind has solved problems while he slept. Anyone who digs deeply into earth history (no pun intended) and reads widely will recognize the great thinking that Dr. Daly has published on the origin of the earth and many of its features. In my own teaching career, I have stressed the importance of random thought and reverie after packing the mind with all the facts available on any problem. It works; try it!

Havelock Ellis maintained that "The hardest facts are the facts of emotion." Emotion guides reverie and dreams in unscheduled ways to arrive at subjective solutions of difficult riddles. It is said that the most successful exploration geologist is the one who can discover gold or oil with the fewest facts. Hard hats would call that "using your gut feelings." Every successful man I have ever known spends time in quiet contemplation. Plato said, "We become what we contemplate," and Jesus Christ uttered the same thought: "As a man thinketh, so is he." One can't contemplate while talking, nor can one learn much by talking, so it's best to listen, for every person one meets can be a valuable source of ideas for this grist-mill we call the contemplative mind. Menander (343-242 B.C.), an ancient philosopher, averred that "Nothing is more useful than silence."

The value of considering reverie and dreams as tools

for the advancement of oneself was not agreeable according to E. Stanley Jones: "It's easier to act your way into thinking than to think your way into acting." I believe such activity follows that similar concept from William James about doing something that you detest which must be done, and by so doing you can begin to enjoy it as a chore well completed. Some thought and resolve must precede the action, which may indeed provide thoughtful conclusions. The prior example is an admonition to writers to write every day for several hours no matter what gibberish comes to mind. The mind can soon be electrified into producing valuable writing. What looks like an enigmatic contradiction to reverie simply proves that there are many ways to approach successful conclusions. Success comes to those who play their strongest hands their way!

Mom and Dad Wengerd

Blood Pressure

Everybody has it; many people worry about it; most of us are descended from ancestors who had it; no one really knows what causes it; it's called blood pressure, and if it is high it is called hypertension. There is great lack of knowledge about it. It's very high in many old people, and if you don't have at least some of it, you don't function well in bed with a willing partner. If it's too high, the doctors say you'll die of a stroke: the blocking or bursting of blood vessels in the brain. Micro strokes cause some violent headaches and may presage a major stroke. After the age of 50, the tendency for high blood pressure is called labile hypertension.

My grandfather and grandmother Miller died of strokes at ages of 72 and 74. My ancient grandmother Wengerd died of it at age 88 in a gargantuan fit of anger. My mother died of it at age 77½ as did my father-in-law at age 90. It is obviously a pervasive disease of old age that can be controlled in middle age by various drugs, beta blockers, natural and artificial diuretics, and a host of other drugs to which the body becomes acclimated. One could die if the drugs are withheld. What really is this pernicious curse of modern socialization?

The causes seem to be numerous, but when I was hit by elevated blood pressure of 180/90, my FAA flight doctor said I'd failed the flight physical, the first time since the age of 19. Causes are different in different people, but the major cause appears to be high body tension caused by worry, oft times by family or business matters which seem insoluble. So I researched the matter during a year of very light diuretic intake, which thins the blood, or at least takes extra water out of the body. Family emergencies disappeared. Diuretics are foreign substances, and after a year my blood pressure stabilized at 140/70 (diastolic over systolic) without any medication. Blood pressure is lower in the morning than in the afternoon. Temporary higher blood pressure is caused by many things: overabundant vitamins, anger, coffee, tea, alcohol, and sex, driving in heavy traffic, the sight of a pretty girl (or the nurse taking one's blood pressure) smoking, lack of exercise, being excessively overweight, bad dreams, and only God knows how many other conditions. I venture all are controllable if one is strong.

A bad life situation can be corrected, solved by death of a problem, a sadness overcome. A supple body stretched vigorously for 60 seconds at least three times each day, deep breathing consciously done many times a day and night and one can will one's blood pressure to normal. I know. I've done it, and I continue to do it.

So You've Quit Smoking

There's only one real danger when you quit smoking: You become an adversary of all smokers to the point of being tiresome. Of course nicotine coursing in one's veins cuts hunger, so there's more than a trifling possibility that you'll gain weight. Why quit? Lin Yutang said he became so out of sorts and unhappy that he had to start smoking again just to get along with himself—not quite in those words, but you get what he meant.

There are many ways to quit: ease off the coffin nails one per day; chew on gum, fingernails, mints, and toothpicks; tell all your friends, neighbors, acquaintances so you'll be ashamed to start again. Mark Twain said he quit many times and it was easy. The scary way is to have your doctor tell you that you'll live only three more months. You could get yourself marooned on a desert or Arctic island where your only choice would be to chew palm roots or muskeg. There are certain chemicals and drugs that appear to be effective if you can live through the ordeal. Or one day you wake up with cotton in your mouth and a hacking cough that raises your blood pressure 50 points, and you look in the mirror, not recognizing yourself as that living youth of past years, and you throw away all your cigarettes, cigars, pipes, and pipe tobacco. By afternoon you try snuff (snooze to some people) or, worse yet, you pack your mouth with chewing tobacco, then you drool all over your shirt and through ineptness swallow the juice because you're inside the house and you don't want your wife to see you with a chaw in your mouth. What to do as you swear to give up tobacco all together?

I started to smoke a pipe in graduate school. A roommate was hooked on Heine pipe tobacco; graduate work was heavy, relaxation not possible. Down to Harvard Square to S.S. Pierce Tobacco Shop which advertised a pack of George Washington rough cut and a pipe all in one package for 25 cents. Remember that was during a depression so deep that tuition at Harvard was only $450.00 per half year, and one could rent a room in a rathouse for $4.00 a week, a toilet down the hall or upstairs. Smoking didn't go well for awhile, just as two or three cigarettes tried when I was in high school tasted like dirt. But by shear courage and effort, I became a devotee of the pipe and Edgeworth or Bond Street tobacco. This led to tampers, reamers, pipe cleaners, and liquids to deodorize the pipes (of which I bought some twenty) special pipe lighters because I smoked more matches than tobacco. With all that paraphernalia, I surely had no pockets in which to carry the necessities every man must carry—and for which a woman carries a purse the size of a gunny sack. Bronchitis and five sieges of bronchial pneumonia throughout a period of 44 years led to the realization that smoking a pipe, even though I didn't intentionally inhale, was killing me. Then, too, smoking became unpopular; dirty looks plus "Put that thing out or get out of here" were standard reactions from people other than my patient wife. She enjoyed the odor of a sweet pipe, and, even though she detested cigars, never forbade me to smoke, even in bed. Then one day, now many years ago, throat raw, clothes astench with acrid smoke even dry cleaning couldn't remove, I quit cold turkey. For three weeks I missed the pleasant after-dinner pipe. For three years I dreamed I was smoking pipes and cigars and, through the haze of waking up, cursed my weakness in giving in to that pernicious habit. Coming fully awake, I realized those smoking reveries were dreams, and I hadn't smoked after all. Elation! Then I became allergic to any and all tobacco smoke, yet I could never revile my friends or ask them not to smoke—except NO ONE smokes in our house, my office, my shop, or in our cars or boat! I invite them out!

The Old Diseases

Whatever happened to some of the old diseases one heard about or saw people suffer from many years ago? I remember a cousin of mine having her tonsils and adenoids cut out. One never hears about adenoids anymore. A sore head was obviously a headache (one of the oldest diseases of man) with many causes: mainly stress or too much booze. The old term bellyache now is called stomach distress; rednecks, however, still call it a bellyache. How about lung fever, a term used for both pneumonia and tuberculosis many years ago. I'll venture not a person under the age of 60 has heard of the disease we used to call erysipilas, a red skin rash. It takes real research to find its modern name. We still have the old term backache, one of the most common diseases of humans today, mainly because our upright body posture has not evolved as fast as our mental capacity. Man, long before he was man, stood up out on the grassy velds of Africa because he ate too much in the trees and got too heavy. By standing up he could see over the tall grass which hid him from voracious animals. His body, unable to walk, run or swim as fast as many dangerous animals, began to develop a curved spine: the sigmoid arch. His feet, bearing weight formerly distributed on four limbs, began to evolve to the condition where the toes were not opposable as are the fingers, and a heavy club-like foot developed a ball heel. Right there the Achilles tendon began to give trouble: the old fashioned footache.

Standing up, man found in front of him two limbs with fingers. What to do? Make something! That was the giant engenderment of intelligence beyond the simple grasping of food. On came rheumatism, first in the hands, where, as in other parts of our stressed bodies, it is now called arthritis—from the dangerous rheumatoid arthritis to the lesser types due to calcium growth in joints. No one talks of rheumatism nowadays. And who uses the term catarrh anymore? It's now called sinusitis. And the crudest word yet devised for a disease as old as man and woman: the clap! How about "blue ball"? There is a cure for this disease, but the old words, still used in bar rooms and pool halls, are for the disease gonorrhea. A "wasting disease" that finally affected the brain is now known as syphilis. In the future will AIDS be an old name for a scourge to be renamed with a less horrible name because a cure is found?

One of the worst scourges of mankind is the mosquito bite, which is not a bite at all but a proboscis sting. Yet the old term holds. When God made mosquitos, it was not the only bad mistake made by that so-called Heavenly Administrator. Owing to the penchant of humans to eat the wrong things, not to stay mobile enough, and to drink the wrong fluids, the human body is cursed with many ills, some self-inflicted, others caused by bacteria, germs and all kinds of parasitic microbes and critters. Most of the old diseases have vanished into fancy medical terminology little understood by any of us—least of all by medical doctors.

The Long Way Back

It's the seventh notch that tells the tale, that last notch on a belt that signifies weight loss has been maintained. Each notch showing greater girth means five pounds gained; thus every inch less on a new belt means five pounds lost. Obviously, for psychological reasons, buy a big belt and start in. It's true that every pound lost before the age of 40 is easy; every pound gained after the age of 40 takes three times as long to get rid of. Why is that so? Less physical activity, more sitting than walking, more walking than jogging—which in itself is the world's worst exercise. That's almost a straight quote from my late friend David Hyde, a Doctor Cureton graduate of Springfield, the well-known physical education university in Massachusetts.

One evening, home from graduate school, I was cleaning my leather belt—then a svelte 34 inches in length—when my dad said, "How about cleaning my belt with that saddle soap, while you're at it." Now my dad, 5' 10" tall, shoulders seemingly a yard wide, wide hips, great legs, and strong arms, weighed about 300 pounds. Massive a man as he was, he was quick as a cat. For years as a young man, he was the champion side-hold wrestler in the state of Ohio; no one ever threw him. I saw him throw my 200 pound brother to the ground—the difference in age was 20 years. All of us boys in the family, along with Dad in his younger days, were roofers. This occupation causes one to develop very thick stomach muscles. What happens to thick stomach muscles as a roofer ages? Any roofer, no matter how thin? They sag! As I cleaned Dad's belt, it seemed tremendously long, so I measured it: 62 inches!

Like father like son is true; the weight battle is continuous in any healthy active male who has been an athlete. Diet plus exercise fight a losing battle with an inherited family characteristic. My churlish friends say rather pointedly, "You eat too much." What I'd like to know is why do so many of my skinny friends die of heart attacks at the age of 50 to 60? When my dad died near the age of 80, he weighed 350 pounds, plus or minus. He never would allow himself to be weighed. He drank large quantities of water—in his youth large quantities of beer—and he ate like a horse. I remember his efforts to reduce, and they were wildly unsuccessful because my mother was a great cook.

The loss of weight is more difficult as the aging process goes on. Crash diets don't work. Simply eating less doesn't work. It requires two added considerations: exercise and control of WHAT one eats. It's an old saw among M.D. physiologists: "It's unhealthy to lose and gain weight after the age of 50." Nonsense! It's easy to gain 20 pounds in two months; it's quite difficult to lose the same 20 pounds in four months. I have aging friends whose weight hasn't changed five pounds plus or minus since their high school days. All of them eat like eating is going out of style. What I would like to know is why four of them in the last five years have died before the age of 70? There must be some kind of vitality factor in the gene pool of some overweight people that allows them to live a long time. My thin friends do point out that one doesn't see many old, fat men. Did it even occur to those friends that older overweight men might not look old?

Dieting is a heartbreak situation. My kid brother is six feet tall, weighs 260 pounds, is as powerful as an ox at 72, and is happy. Once in awhile he goes on a really tough diet. Daily he takes three 300 milligram ferrous sulfate tablets, three eight ounce glasses of whole milk, one one-a-day vitamin tablet, three cups of black coffee, and NO alcohol. In three weeks he loses 30 pounds, looks and feels like a million but is unhappy. My own doctor, with whom I constantly argue about physiology—inasmuch as I was once a pre-med student before I learned that I didn't like the sight of blood—tells me my weight is exactly right for a man 6'2" tall. Until he went on a diet and lost 70 pounds, he never mentioned my weight. Though he was so thin he looked awful, he began to harp at me to go on a strict diet.

My brother Wilmer and me
April 7, 1993

James McKenta observed, "It is right to be contented with what we have, but not with what we are," so it's no more than two eggs a week (although three of my doctor friends agree that the medical profession can't yet make up its mind about the so-called "safe" level of cholesterol!), no or little salt, fewer than 30 grams of fat per day, a balance of carbohydrates and protein, lots of fruits and vegetables, some chicken or fish, no red meat, plenty of fiber, and at least three hours of sustained exercise a week to get that heart rate up to 150 beats per minute. For every doctor who will tell you no or little alcohol, there is another who will tell you that two cocktails a day after age 40 are excellent. The fact which the medical doctors routinely overlook is that everyone has a different chemical make-up, a different psychological bent, a different electrical system, and a different outlook on life. One thing is certain: smoking is unhealthy and unnecessary!

Grudgingly, a doctor will say it's okay to take vitamins to bolster our deficient American diets; others will vehemently deny that extra vitamins are necessary; some will say they do no real harm; still others say that vitamins are useless. They can't make up their minds because they don't know, and to cover up their ignorance, they make bold inaccurate statements. With proper exercise and a carefully planned adequate diet, I have found the following daily vitamin regimen to be excellent FOR ME: 400 units vitamin E, two grams vitamin C, one capsule Poten B-50, one 25 to 50 milligram zinc tablet, one 50 milligram potassium tablet, one ounce of lecithin granules or one 19 grain lecithin capsule, one 20 minim wheat germ oil capsule. Once per week a 25,000 unit vitamin A capsule and one 10,000 unit vitamin D capsule, and one to three Omega 3 fish oil capsules per week. People who are on the one-a-day vitamin regimen do not realize its vast inadequacy. A thin friend, who had a serious stroke and has since died, told me he was convinced that if he had used lecithin granules regularly he would never have had a stroke. He maintained also that the only oils one should use are safflower, sunflower, corn, and canola.

What then is the answer on the long way back from an overweight condition? There is no one answer, no one combination of answers for everyone. All must find the amounts and types of herbs, foods, vitamins, oils, meats, vegetables, medications, eggs, fibers, beverages, and exercises that allow each individual to function best physically and mentally as we enjoy life. Sachel Paige maintained that age is a matter of mind over matter; if you don't mind, it doesn't matter. Someone once said, "Not all are privileged to grow older gracefully in full command of all their faculties." The greatest condition is to stay upright, mobile, healthy, smart, sexually active, friendly with all humans, contented, and have a joyful outlook on life at whatever weight feels good!

Grandma Catherine Miller
with great-granddaughter Norma Jean

Old Ladies

Old ladies are a very special part of our society, and they are very interesting, partly because there are so many of them. They've lived a long time, had many experiences, do not fear the truth, and are unusually short of temper with younger people who act stupidly. I admire most of them, partly because they're survivors. They make up well over three-quarters of the people who are over 80 years of age. Of course, those between 60 and 80 are interesting for many reasons other than their self-assured wisdom. Strangely, I never thought of my mother or my mother-in-law as being old ladies, for their relationship to me began what seems like eons ago. Both are long gone, and their love and guidance were valued parts of my education. They were what advanced maturity should be like: calm, peaceful, no more rushing around, kind, cheerful, with soothing low bell-like voices. They grew older with grace and wisdom and were observant of the vast changes in our society between the ages of horse and buggy and space craft; they lived through three wars—Spanish-American and World Wars I and II plus our sad adventures in Korea and Vietnam. Both saw two sons and sons-in-law go off to war and worried about a world that could tear families apart.

So much for the encomiums; now for some critiques of old ladies totally unlike my mother and mother-in-law. Why do so many old ladies cackle? Their voices come through their noses from the backs of their tongues. Their laughter sounds like old hens cackling in a barnyard when a rooster struts by. Why do so many dye beautiful white hair a hideous blue? Why do some of them dress like teenagers? I don't ask for somber dress. Colorful dress is great; a flounce to a skirt on a racy old lady is a thing to behold. And if their eyes glitter, watch out! A mature spirit doesn't need to look like an immature clothes rack in short shorts or hip length dresses. And then there are the fat ones who wear shorts: better to put on a muumuu—or a tent. After all, what is not seen can be most fetching. And why the heavy makeup with blue stuff on their eyes? What's wrong with well-cared for wrinkles? Who says grandmothers and, for that matter, great-grandmothers, should dress in somber puritanical black or gray and sit at home twiddling their thumbs? God bless them all; they're great!

Never Run Downhill

Our home is just across the street from the Bernalillo County Health Center, between the UNM Medical School and the UNM Law School. To the north and west lies a nine-hole golf course around which there's a two-mile running track. One of the last large pieces of clear real estate on the edge of the East Mesa above downtown Albuquerque, this area is popular for walkers, joggers, runners, and bicyclists. Dogs are not allowed, but the UNM cops drive by twice per day, and there's a sour-puss of an old man who patrols in a golf cart to mess everybody up. The two mile course is uphill, down dale, along flats, with water piped to five or six drinking fountains for golfers and joggers. All in all it's a haven in the city.

Joggers are a strange lot. They run along glancing at their wrist watches every few minutes with serious grouchy looks, listening to Walkman radios and tape players taped to their belts, earphones drowning out all the birds and other lovely country sounds. With their heavy breathing and sweaty bodies, they do not notice the lovely odors of the desert dotted with cottonwood and elm trees, the gardens of the adjacent large homes on "Pill Hill," and the clear air untrammeled by car fumes. They jog slowly up hills and run pell-mell down the gravelled hills when paths are slippery. Falls are common; scraped and twisted knees and ankles normal. Why do they run down hills? Hills are designed to amble down, to tarry by burrowing owl and prairie dog holes, to watch coveys of quail and the two jack rabbits that inhabit the golf course. There are many kinds of birds—most notable, at least, two pairs of roadrunners, the state bird of New Mexico. There are even a resident prairie dog, several picket pins and dozens of other ground squirrels which grab stray golf balls to secrete in their dens. While others run willy-nilly, unjoyful in these surroundings, I hike along finding such things as a tennis ball after a very heavy wind—the nearest tennis courts being over a mile away—a five dollar bill, miscellaneous clothes, towels, shirts and sometimes a pair of women's panties—little wonder why they may be there, for the area is isolated, quiet, grassy, with a giant culvert at the north end to hide in on a warm evening. Golf balls, yes. Over the 20 years I have been jogging up hills and walking down hills, I've found at least 200 golf balls. Jogging every other day when we're home only a few months of the year, I find at least one every third round and as many as four and five after a heavy Sunday golfing day. I've supplied my brother in Florida and my brother-in-law in Ohio with golf balls for years. Do they ever offer to pay postage? And I never run downhill since I've learned to appreciate my surroundings.

Part Ten

CLOTHES HORSE

Hats
The Wandering Pocket Watch
In Praise of the Seven Button Shirt
Hip Room in Pants
What Happened to Spats?

Hats

There are four things a man considers very personal: his hats, his toothbrush, his underwear, and his shoes. It is a truism as well to note that good cobblers and capable barbers are two of the most important people for any sane man to cherish. But right now the subject is hats. In the West, particularly West Texas, one hears the admonition, "If you ain't got a hat, yer a nothin!" A man's very character can be appraised by the hats he wears and how he wears them. That includes caps and any other kind of headgear. We'll limit this to men, because I've yet to fathom the personality of one of my wife's acquaintances who wears lamp shades for hats. That throws me for a loop. Among workmen, a cap or a hard hat is a badge of hard work, so they wear them indoors, when they're eating and almost everywhere else except in bed. Of course, wearing a hat while making love is simply a no-no.

But I'm off the subject again. Hats and caps are clues to a man. If at a jaunty angle, he's probably a rake. If pulled down straight over the ears, he's likely to be stupid or cold. If the hat is on crooked, the bill or snap brim off to one side—like Gerald Ford too often wears his golf cap—watch out. He's likely to smash you with a golf ball or let his club fly in your direction. Since the introduction of the batting helmet, one can't tell much about the capability of a baseball player by his cap because his head's covered with the helmet while he's batting; but watch how he wears his cap on the field. I'm convinced there's a direct relation between how a ball player wears his cap and his batting average and the number of errors he makes. You'll have to do your own analysis. If you're betting on a baseball game, don't forget the dictum which says "Never bet on anything that can talk!"

One can tell a lot about a man who has myriads of

caps and hats from all over the world. I have a cap crocheted by a Mexican Indian in Amecameca, a little village high in the saddle between the Iztaccihuatl and Popocatepetl volcanoes in central Mexico east of Mexico City. Friends tell me it's disreputable, but it's so comfortable—all wool, keeps hair neat and looks like heck. I've even worn it to bed at night in the wintertime (too cold for sex).

My Amecameca Cap

I can't possibly tell you how many hats and caps I have; I've run out of room to hang them all. Every time I go to Mexico, I buy a palm or straw hat. They wear out fast, and when I leave one in my shop, the cockroaches eat it up. Also rain ruins them, so I have to buy new ones every time I go South. One time, years ago, when my number one grandson was a year old, he looked at my assemblage of headgear strung up on a wall. He said, "Hat, hat, hat"—probably the first time he had said any words. So my hats have had a hand in his highly developed loquacity.

There's another time when it should be absolutely forbidden to wear a hat, and that's when you're brushing your teeth! You'd be amazed how many low-brows put on their hats right after getting out of bed, forget to wash the sleep out of their eyes, wolf down breakfast, and then brush their teeth with their hats on. But I guess one can do as one pleases in one's own home.

Cowboy Sherm
Oklahoma
August 1939

The cowboy hat is one of the most popular types for men and women. It keeps the sun controllable, the rain off one's head, and you can hide behind it if you're caught skinny-dipping in a cattle tank. This happened to a friend of mine when I was a weekend cowboy in western Oklahoma. Hot and dusty after herding cattle from one pasture to another, we stripped off and plunged into the shallow earthen cattle tank. A lady tourist stopped to ask us where the Baptist church was in Woodward, Oklahoma. Now the cattle tank was so shallow that when he stood up to point—as naked as a jay bird—I had to give him my hat to cover certain parts. Why did I have my hat on while swimming? I always wear my hat in a cattle tank. He learned his lesson so well he now wears his hat when he goes swimming in the municipal pool!

Main Street
Woodward, Oklahoma
1937

It's interesting to survey all the kinds of materials hats are made of: rabbit fur, felt, cloth, straw, palm fronds, cord, willow reeds, and God only knows what else. Hat makers copy each other. Some years ago I bought an Irish woolen hat in Pendleton, Oregon, just before the round-up. It was made in Pendleton from Oregon sheep wool. I put it with my collection until 1990 when we were in Scotland and I bought another Irish woolen hat. (Don't the Irish make those floppy hats anymore?) So now I have two almost identical, very comfortable woolen hats. In 1960 I bought a British shooters cap, and when I wear it in New Mexico, derision is my lot. They think I'm putting on the dog. The best hat I ever had was a dark blue Homberg. A Tyrolean hat with feather intrigued me in Interlaken, Switzerland, but the canny Swiss had built almost immediate obsolescence into that hat. I lost the feather, the sweat band was paper or plastic, and the brim fell off.

So much for hats. I simply can't take reams of paper to tell you all the stories about my hats. Next time I write about hats, it will be about my father's hats—a rousing story!

1932 High School Graduates
I am in the center of the bottom row

The Wandering Watch Pocket

Whatever happened to men's watch pockets? Of course everyone over the age of 40 remembers the vest pockets with the dangling gold chain, a watch on one end and a coin, rabbit's foot, nail clipper, or gold thing-a-ma-bob on the other with the chain threaded through a vest button hole. But that was for the tycoon. What about people who didn't wear vests? I remember overalls, the working man's uniform of decades ago before coveralls were invented. In a breast pocket was space for an Ingersoll or Westclox dollar watch—in fact, there were many types of inexpensive watches available. Remember the railroad man with his expansive vest and massive gold or silver chain anchoring a heavy railroad watch?

But what has happened to the watch pocket over the years? It's moved several times! When pants came without bibs, the watch pocket was put inside the belt line on the right side. If suspenders were worn, the watch pocket snuggled between the buttons of braces, then called galluses in a working man's pants, or suspenders in dress trousers. Can such a pair of pants be found today? Not in pants one can afford. Tailors will put one in at an exorbitant cost, but they'll tell you it creates an unsightly bulge in a man's trousers. Besides, the watch pocket is difficult to anchor properly.

So NOW where has the watch pocket migrated? Look on the right side of your sport coat pocket. Inside that outside pocket is a small neat pocket for a watch. No sightly chain to fool with or impress your neighbor, such as was necessary to anchor your watch in the small pocket inside the belt line. Thin-line watches now available are expensive, but how many men do you know who wear a pocket watch? They have almost been driven out of existence by wristwatches, and particularly by that travesty, the digital watch, which has no second hand except for a blinking colon.

The gold pocket watch is now so valuable that it can be a collector's item. I have one: a heavy 24 karat gold Elgin with a highly visible second hand, a tick that is music to the ears, and a weight that makes one feel like a millionaire, even though I seldom wear it. It was given to me by my dad when I was eight years old. I prefer a Westclox-type which one could buy years ago for a dollar or even 98 cents. The one I bought with a second hand in Denver in 1972 cost $12.95 plus tax, but recently I found one without a second hand for $8.95. I had one some years ago which my wife put through the washing machine. It worked fine, but after the second time, it gave up the ghost. So much for forgetting a pocket watch in a pants watch pocket! I keep the old ones for my grandsons to take apart, which they do with gusto, making tops out of the spindled, cogged wheels. Lastly, to my amazement, I found a watch pocket in a new khaki jacket the other day. Where? On the left side below the outside pocket. How time moves!

In Praise of the Seven Button Shirt

There is nothing quite as exasperating as a shirt with only six buttons down the front, and the six include the button at the collar! Buy a shirt—sport, dress or knockabout—with only six buttons from a mail order house and you've got a disaster. First, a gap will appear above your belt, and a pink belly looking out above a man's pants is no thing of beauty. Second, some cheap shirts even omit the collar button, another travesty on justice in a man's world of clothing.

I'm convinced that people who design shirts never wear any. So why do dress shirts have only one breast pocket? To save money? A man needs lots of pockets. Now the Guayabara shirt, ideally designed, has pockets below the breast line—more toward the tummy line—so why not have four pockets in front? The Mexican or Guatamalan who designs such a shirt will steal the market from London, Paris and New York—also Peoria, Illinois.

And another thing. Why do shirts have those ridiculous tails so you can't wear them outside your pants when you wear a T-shirt with a classy design on the front? Cut the bottom of a shirt straight around and you have a shirt-coat for summer formal wear. Of course this would frustrate the female crowd of younger than thirty who simply consider it *de riguer* to wear men's shirts with tails flying above slack tops. That's not too bad if it's a six-button shirt and the girl leaves the top two buttons and the bottom three buttons unbuttoned, has voluptuous boobs and a neat, flat tummy.

There's another disgusting thing about most men's shirts: why, on the button-hole side where there's a double thickness, isn't the flap sewn down so it doesn't get in the way of buttoning the shirt? It's ridiculous to see a man with the flap caught and covering the button he's struggled to fasten but the flap got in the way. When he comes to unbutton his shirt, he tears the button off trying to get out of it. I'd like to know also why women's shirts button the opposite way of men's shirts? To assuage the ultra-feminists, I should have reversed that question, I suppose.

Short sleeve versus long sleeve, knit versus broadcloth, pearl buttons versus plastic, there isn't a shirt in the world that has its buttons sewn on with thread stronger than a spider web. Why? And why do so many buttons have only two holes? Why do so many get lost? Because they're not sewn on with tough fish-line casting thread, which even comes in colors to match the buttons.

I offer all the shirt makers of the world these splendid suggestions free of charge as I praise the seven-button shirt. In fact, why not an eight-button shirt?

Hip Room in Pants

What in the world ever happened to hip room in pants? The Dutch have the right idea and always have had. Make pantaloons wide and stretchy with room for hips of any size. In the old days of tight pants with cuffs at the knees, white powdered wigs and ruffled shirts, what would a man do if he wanted to carry stuff in his pockets? Wear a great coat! Then came freedom. Somewhere between Thomas Jefferson and Andy Jackson men became smart and rid themselves of that foppish European look. They began to wear long pants with four spacious pockets and waistbands large enough to wear wide belts and wide-hipped trousers. Then someone thought to design galluses to hold up the pants. Finally belt loops won out with ever wider belts. These mimicked the sashes that were so flashy and devilish to tie neatly when men wore short, silky, tight pants.

I have no inkling of the part women played in the design of men's pants in the past, but it seems they must have had a hand in making pants easy to get out of—until men became more feminine and began to wear tight pants with belts, probably so that their buns would be shown off better. Of course, when women began to wear pants, they were loose, wide-hipped and comfortable. Then the lithe cowboy was envisioned. Those long slender cowboys of the movies, such as Gary Cooper and John Wayne, were not as slender as you may think. They were big, wide-hipped men who rode a saddle almost as wide as the horse. Most working cowboys in Texas are short, chunky, powerfully-built hombres with wide shoulders, big hips and adequate waists needed to ride the wild range. If you had been there, you would have noticed that Bob Crosby, one of the world's greatest cowboys of his

Bob Crosby

time, was wide, squat and wore pants with lots of hip room. I know because I knew Bob Crosby. Next time you go to a rodeo, notice that the older cowboys, who are the champions, are built like that and that they've got hip room in their pants. You'll never see a real cowboy poured into a pair of pants like women wear Levis nowadays. If they did their paunches would split seams in all directions.

Nowadays many pants bought by men for everyday wear are made in the Far East where people are small, and, right there, hip room is a thing of the past. There is a ratio of belt size, hip width, bun room, pant-leg diameter, and leg length that foreign pants makers find difficult to duplicate for Americans. So, if you want to buy comfortable pants, buy

American-made workman's pants; yes, even overalls or coveralls. Why not wear a white shirt with bow tie, neat coveralls with a silk stripe down the legs, and a tuxedo coat as formalwear? Men, declare your freedom. Strike for more hip room in pants. You can even call them trousers if you must! And as for cuffs on pants, good riddance!

Wooster friends—
Henry Lee and Paul Yee

What Happened to Spats?

Anyone who has lived a long time can't help but be amazed at the evolution in men's wear. The greatest changes in America involved the change from dandified silken knee britches to long pants and from brocaded jackets to suit coats between the time of Thomas Jefferson and Andrew Jackson. But in the last 60 years, there have been interesting changes of a more minor nature. Take spats. In the cold Northeast, spats were not just fancy foot wear for parties and dances—as we wore them in college—but were worn to keep ankles warm. Winter wear was heavy. Boots for dress wear were not in style, as they are now. And oxfords with spats were not just for dandies and jelly beans, as we called the rich boys in college. Anyone seen spats lately? Even the diplomats have quit such dated wear. They've become part of period costumes!

How about arctics with the slot and tongue buckle that tightened those rubberized cloth uppers around one's ankles? Of course such are worn in the cold Arctic and by men and women who work outdoors. Such clod hoppers create laughs today, yet we wear them in the outback of the West where snow is battled every winter. Wellington boots of leather and corfam now grace more feet. In the United Kingdom, Wellingtons are the rubber half boots and knee-length boots worn for watery jobs. Fishermen on the docks and on fishing boats wear them with heavy socks, whereas the Arctics are worn over shoes.

What happened to garters, sometimes called sock suspenders? Their demise was really welcomed with the appearance of knit socks which stay up without the need for these circulation-stopping old fashioned tortures. Try to buy a pair, though I don't know why one would. As a pack rat who wore them in the days when socks fell down around one's ankles, I see my last pair once in a while in a what-not drawer. I should throw them out, as they are not a source of nostalgic memory. And the colors—bright red, sickly green, striped black and white. The worst view of them was when a fat man in shorts wore them out of force of habit. Ridiculous!

Arm bands, known as sleeve holders or arm garters, were worn to shorten too-long sleeves. These were drab or colorful items of style. It was obvious that one could judge a man's character by his sleeve holders. Of course, worn with suit coats, one could only guess. You'd be surprised to know that many men wore matching garters and sleeve holders. Makes one wonder if their underwear also matched their garters.

In the dandified past of men who wanted to be in style, detachable collars and cuffs were worn regularly by business men. These required cuff links with collar studs in front and back to fasten the collar to the shirt. Before the days of automatic washing machines and no-iron shirts with attached collars, many a man simply changed collars and wore shirts until they'd stand by themselves in a corner. After all, taking a bath before there were electric hot water heaters, running water and showers was a major weekly undertaking.

Shoes. Foreigners say they can always identify Americans and Canadians by their shoes. There was a time when shoes were only brown or black, and a man who wore white shoes was considered to be a gangster. One can tell much about a man by looking at his shoes: elevator shoes typify the vain man, alligator shoes and boots are typical of the kind of man who drives a Mercedes. The appearance of shoes made in China, Korea or Hong Kong costing far less than the excellent footwear formerly made in New England has changed the rules of judging a man by his shoes. The old-time canvas shoes have been replaced by styled sneakers, tennis shoes, running and walking sports shoes, fancy Adidas types high and low. Woody Allen, always the iconoclast, has been seen wearing basketball shoes with his tuxedo.

Advertisers assert that clothes make the man. I

especially avoid comments about women's clothes, only noting that "the less the better." Fair warning to car salesmen who tend to set prices on cars by the clothes a would-be customer is wearing. I know one man who dressed like a bum in clean but old- fashioned clothes. He was treated snottily by a salesman, so the man bought the car agency and fired the salesman! The clothes a man wears tell you little about his intelligence, the amount of money he has, his character, or his morals. One of the richest men in the state of Ohio wears old hand-me-down leisure suits, with garters, sleeve holders and galluses bought at a Goodwill Store! He's the owner of 27 corporations.

A boat ride in
Singapore Harbor

Part Eleven

WHY GO OVERSEAS?

So You Want to Go to Europe
Backcasts of Europe
Tracks in the Sea
The Seaman

So You Want to Go to Europe

Many guide books have been written on travel and living in Europe. Except for descriptions of ancient buildings, battle sites, other historic places, and natural features, such guidebooks are usually out of date by the time they're printed. The American tourist is seldom also a traveler. He thinks he must have reservations wherever he goes. Moving around with phrase books, he attempts to get to know the local citizenry, calling them natives, yet he chooses to move from tourist hotel to tourist hotel, uses only guided tours, and tries to "do" Europe in two weeks. Seldom an expert in language, cursing the local money as "funny money," he stumbles around and, at times, becomes the Ugly American. Unable to recall his high school French, he shouts a gibberish of English in what he thinks is a French accent and becomes convinced that all the "locals" are stupid. I even heard a woman tourist berate a shop-keeper one time: "Why don't you people learn to speak English?" The shop-keeper agreed, not knowing what she had said.

Too few Americans study the countries they wish to visit. They refuse to learn even the most basic phrases of social intercourse and are confused by exchange rates and how and where to exchange money most efficiently. I was in Europe one time when the dollar was of such low value that no one would take dollar bills for tips. I became angry at our government, not at the locals.

There are many easy ways to travel in Europe. Their trains and buses are efficient and on time. Forget taxis; taxi drivers are the same the world over: rude and too expensive. Yet when one thinks of gasoline—petrol, *l'essence*—at $3.00 or more an imperial gallon, no wonder taxis are expensive. Imported cars are duty taxed from 50 to 150 percent. In

Finland it is 120 percent and in French New Caledonia 150 percent of their original cost.

Most European highways are great. In 1960 my family and I drove through all of Western Europe, except Austria and Portugal—7000 miles in three months—in a Volkswagon Microbus, with a top rack carrying 27 pieces of luggage for six people. We had reservations in just two places: in Rotterdam at the Hotel Continental and in the Dordogne Valley of France at the Cro-Magnon Hotel in Les Eyzies, where Florence's parents were to meet us. We didn't even have reservations in a hostelry in Copenhagen where we were to attend meetings of the International Geologic Congress. We carried a hefty letter of credit from our bank and never used it. Travelers checks in denominations of $20, $50, and $100 are easily exchanged for local currency, but be prepared to suffer a mild kind of embezzlement on every exchange. Banks are banks the world over, and some of the most costly to deal with are in Europe. Don't forget your passport; it's necessary for all kinds of negotiations.

Although it's dangerous to lump all Europeans in one mold—for a Norwegian is as different from a Spaniard as a North Dakotan is from a Downeaster—there is one common denominator: they are generally jealous of our freedom. They will tell you that Americans are too open, too trusting and have too much money. Everywhere, particularly in large cities, pickpockets are abundant, especially at night and in crowds. Never leave valuables in hotel rooms; check them into the hotel safe. Or, if you're a big, husky person, carry them in a small underarm bag over the shoulder under a coat. Look like you're a menace—it pays off. If anybody, especially two people, seem to be following you, they are! Duck into a store or doorway, or walk straight at them with a sneer on your face! All in all, we found the Europeans to be very friendly, except the Parisians. All country people want to know where you come from, what you do, who you are. Answer them, and ask questions. It may be their country, but it's everyone's world!

Captain Sherman A. Wengerd

Backcasts of Europe

There are many reasons for going to the old country. There is much to see and many interesting people to visit. We try to find out the kinds of areas where our ancestors lived. We attempt to understand the geography that gave rise to fiefdoms, kings, the powerful Catholic Church, knights and serfs, castles and hovels. And by reasoning and visiting, we try to visualize why our forebears left kith and kin and the safety of the known to brave untold dangers in new lands across the wide western ocean.

But when we arrive, with some ability in language, we find people strangers to Americans. We soon recognize that the people there today are descendants of the people who stayed behind and that they have also changed radically since our ancestors emigrated between 1600 and 1800. We finally realize that to truly find out who we are and what we have come from, we must study history with great care and imagination.

Not that the countries of Europe are a disappointment today. Here are the things I found enjoyable on our first trip in 1960 when my wife and I took our four children to Europe on the *Maasdam*, landing in Rotterdam.

There are windmills as in fairy tales; castles with moats and portcullises; silver leaves on green olive trees cascading down across red clay hills; vast vineyards in Italy, Spain, France, and Germany; hay rows in the low, wet countries; small haystacks in Spain and France; conic trees—dark gray, unstripped with red underbark where the thick cork is skinned off; a Spanish wine bottle in San Sebastian for more pesetas than the cost of wine drawn out of casks in caves in limestone hills; the churches of Voldoy and Molde in Norway with their great sonorous organs and the ship model suspended from the high vaulted ceilings; the sky-piercing cathedral spire over 300 feet tall in Ulm, Germany, the tallest in Europe; the threshers who work with crude scythes in Italy and Spain, their grain-loaded carts drawn by donkeys or oxen, these animals circling and trampling the grain on hard, bare ground to separate wheat, chaff and straw (when it's windy the peasants throw the grain into the air to separate chaff from grain); the women and whole families who work in the fields; travelers having lunches along the road, by the fence rows and in the harvested fields on weekends; the incredible number of bicycles, especially in the Low Countries; and the canals with loaded barges drawn by mules and oxen. One notes the merging of languages such as German and French in the Alsace-Lorraine, Celtic and French in Brittany, and Celtic, German and Spanish in northwestern Spain.

There is a distinct individuality among the French, with their refusal to listen to English, and their mockery of those who speak French unlike the way it is spoken in Paris. One feels the strong antipathy the Teutonic Germans and the Berliners have for the Bavarians—and the Bavarians care not at all what the Teutons think of their *nieder deutsch*; in fact, the southern Germans and the Austrians, fun-loving and musical as they are, find the Teutons insufferable. Neither kind of German gets along with the French. The French exhibit a fantastic effort to keep their language pure, when it is, in fact, one of the most impure polyglots in the world, mainly because they don't realize it is truly a mixture of Latin, Greek, Slavic, and Old German.

The Europeans are a happy people; they know how to sing, dance, drink, and make unashamed love. Of course the German Lederhosen crowd that won't tip makes Italians and Spanish waiters want to shoot them on the spot. One finds the Baedeker and Michelin guidebooks woefully out of date, especially about hotels, prices, meals, and newly revised highways. As one travels and talks to the urban people, one senses a strong socialism, quite different from the rugged individualism of the country people.

It is no wonder so many people want to leave Europe, with its lack of opportunities. But look at all who stay and make the best of a crowded situation. Once you get to know people well in Europe, they are critical, and many will tell you, "But you are so occupied with things. We like to think that we are occupied with things of the mind." When they are friendly enough, one can retort with some orderly comment such as, "You really have many things because you are so happily disorganized with the things of the mind and body. You desire to have the government take care of you; you aren't efficient enough to create the things you so admire about America. Our capitalistic system engenders the spirit to make advances that comfort the body and really give the mind and body time to relax." In the more socialistic countries, the suicide rate is high, for the ambitious are so regulated that progress is very difficult. In fact, many wish that a material culture would be developed for them by the government.

Our concept in America seems so antithetical to many people of Europe, to wit: the mediocre should, by their own choice or ineffectuality, lead a mediocre life. All this was impressed upon me in 1960, and I realize that since then great changes have taken place almost daily. The changes are revolutionary. One of our Danish friends, a lawyer, explained that Danish socialism guarantees that all will be at the same level economically. When I retorted that such a level was bound to be a very low level of mediocrity, he was shocked. Our argument carried over to labor unions (by government fiat very powerful): stifling change, protecting the inefficient, and causing large businesses to move elsewhere. Super protection of monopolistic business means that failure is rare and, obviously, so is great success, but, without failure of some, there cannot be progress for many.

Socialism and its attendant false security in Europe may be due to inevitable overcrowding in small countries. Confiscatory taxes discourage investment in new ventures which are already stifled by government laws inimical to competition. Competition is the very heart of American capitalism. Perhaps our almost wide open system is in part due to the vagabond fluidity of Americans: "follow the jobs, move or stagnate." This may be the result of our country's size and the fact that anyone with imagination, unafraid to take a chance and to work hard, can succeed, but there is no guarantee against failure. After all, most Americans are descended from people with enough guts to migrate. The European of 200 to 300 years ago was restless; he may have been less successful than his peers in his own country who didn't want to or didn't need to take chances and move across the sea. Thus a natural selection took place that populated early America with British, Scotch, Welsh, Germans, a few French, a great number of Swiss (my background), Norwegians, Swedes, Danes, and an earlier flood of Spanish through Mexico. The later floods of Irish, Italians and Mexicans came more because of incipient starvation than a search for religious and economic freedom.

What started this vehement spiel? Several long visits to Europe and talks with many people. The more one travels, the more one loves America despite all its problems. I've been to Europe quite a number of times as a geologist and as an observant traveler. Upon my return home, I always marvel at American drug stores, shopping malls, Sears, Montgomery Ward, Wal-Mart, K-Mart, and similar great stores, auto stores, restaurants, and I thank anyone who had anything to do with my ancestors coming here. Simone Signoret, that lovely, Oscar-winning French actress, said this: "But even today I marvel when I eat an orange!"

Taking a ferry from Newfoundland to Labrador
1994

What do you *do* on a freighter?

Answer—
Write a book!

Tracks in the Sea

They stream behind ships like a broad road, flanked by ever-widening bow waves, the center churned up briefly, then calming with a sky-glinted sheen. These are the evanescent tracks in the everlasting sea. They go from straight to curved as ships turn to new courses, and when a ship is dead in the water, those tracks soon disappear. Where do they go? Are there still waves circling the earth from the sterns of Polynesian outriggers or Phoenician prows of ages ago? Do ships leave tracks, or are these like the holes in the sky over Floyd Bennett Field that fill up after thousands of airplanes have ceased flying there?

The romance of the twenty ships on which we have sailed in many parts of the world—and indeed around the world—cannot be denied. Some of those tracks have been recorded on film, but most are as will-o'-the-wisps of the gentle zephyrs after a spring storm, never to be seen. It's a good thing that they disappear or the sea would become bumpier than it already is by conflict of current and wind. A track in a heavy sea of 80 knot gales and 50 foot waves is a fearsome sight, just as is the periodic to episodic roller or swell that towers over the bridge of the ship to crush it down like a giant sledge hammer. Such a sea afflicted our freighter in the Bering Sea north of the Aleutians. The captain slowed to a crawl to hold her head into that fearsome gale so as not to break the ship's back.

The sea, lit by a brilliant red sunset in the Galapagos Islands, the water cold out of the Antarctic, sent shudders through our 71 foot staysail schooner riding trough and crest of 30 foot waves striking us on a forward quarter. The roll and plunge, pitch and wallow made us hold our plates with one hand and food with the other hand at chowtime, as we galloped eastward along the very equator when the sun *always* sets at precisely 1800—to you land lubbers, that's 6 pm, but I suppose you know that. When it calmed down late at night, sailing was so quiet we couldn't sleep. It takes a bit of rock and roll to sleep well on a ship!

Once in the Sea of Cortez in a sailing folbot, a frail craft, my son, Tim, and I fought a strange current crossing a beam wind that bent our aluminum mast double. Too busy to bend on half sail, we rowed for our lives miles off the Sonoran beaches of Mexico. We were too busy to notice a wake in our two knot effort. Under sail, we had been making six knots—unheard of, literally burning the rubberized canvas off the slender boat. Well, not quite!

During World War II, going west from San Diego to Pearl Harbor on the cruiser-hulled carrier *San Jacinto*, our Naval air squadron was sent to enjoin the war. Our maneuvers comprised a sequence of course changes to thwart Japanese submarines. Guided by two destroyers, this trio of ships, like a mother hen guarded by two chicks, made three coordinated tracks in the sea. With crossing patterns, both forward and aft, sonar searching for enemy subs on the way to the Molokai Channel, we zigzagged along in a deliberate set of random course changes so complex it's a wonder we didn't run down our protectors. No wonder the Japs never got a torpedo shot at this beautiful new carrier on that voyage. We got off at Pearl Harbor to train for our duties of making aerial maps for future attacks on Saipan, Truk, Yap, Iwo Jima, and many other island groups. Using reconnaissance air photographs made by forward fleet units and the Army Air Corps, we made maps and mosaics, while the *San Jacinto* went on. We had hoped to stay on the carrier as a forward echelon air photographic unit to get into the fight, but Com Air Pac said, "No, unless you can shoot guns and drop torpedoes with your TBM aircraft, you can't come along!" The *San Jacinto* made many tracks in the sea on her missions, was eventually badly shot up, and several of the fighting Naval aviators we got to know so well were killed in battle with Japanese carrier aircraft.

After the war, and especially after I retired from teaching, we traveled on many ships: three cruises in the Caribbean, to Antarctica, around South America, through the Panama Canal, to Japan and back, a three continent cruise (Europe to South America to Africa back to Europe), and a cruise from San Diego to Tahiti on a fancy ship called the *Crown Odyssey*. And now around the world on the British freighter *Ivybank* as I write, making many stops between Tahiti and Singapore all through the myriads of island groups I analyzed as an Aircraft Facilities Intelligence Officer in 1942 in the Air Navigation Division of the U.S. Naval Hydrographic Office. The next year—1943—I was chief charting line officer on board Capt. Robert Bartlett's vessel *Morrissey* in the Canadian Arctic and Baffin Island's Frobisher Bay, making charts for ships supplying our Arctic air bases. We made many tracks in the North Atlantic, infested with German submarines. They thought we were innocuous as we relayed weather, made charts, and radioed ship movements. The German sub captains thought our 105 foot schooner unworthy of a torpedo. How little they knew!

En route to Tahiti

The world of ships is an active world, much more benign in peace time than during war, but thousands of ships make these tracks for commerce and for pleasure as we scour the world's oceans to learn more about our marvelous planet. Who has not sailed has missed much of the almost three quarters of our globe covered with 330 million cubic miles of sea water. Try it, you might like to make tracks in the sea!

From the far western Pacific comes the following essay on "Seamen," a contribution in the public domain by an unknown author.

The Seaman

Between the innocence of infancy and the recklessness of adultery comes that unique species of humanity known as a seaman.

Seamen can be found in bars, in arguments, in bed, in debt, and intoxicated. They are tall, short, fat, thin, dark, fair, but never normal. They hate ships' food, chief engineers, writing letters, sailing on Saturdays, and dry ships. They like receiving mail, pay-off day, nude pin-ups, sympathy, complaining, and beer.

A seaman is Sir Galahad in a Japanese brothel, a psychoanalyst with *Reader's Digest* on the table, Don Quixote with a Discharge Book, the savior of mankind with his back teeth awash, Valentino with a fiver in his pocket, and Democracy personified in a Red Chinese prison cell. A seaman's secret ambition is to change places with a shipowner for just one trip, to own a brewery, and to be loved by everyone in the world.

A seaman is a provider in war and a parasite in peace. No one is subject to so much abuse, wrongly accused so often and misunderstood by so many as is a seaman. He has the patience of Job, the honesty of a fool, and a heaven-sent ability to laugh at himself. When he returns home from a long voyage, no one else but a seaman can create such an atmosphere of suspense, excitement and longing as he walks through the door with the magic words on his lips: "Have you got the beer in, then?"

Gloucester, Massachussetts

Part Twelve

FINAL CALL

Pursuit of Ancestors
The Aging Syndrome
The Relentless Tide
Death Anytime

Pursuit of Ancestors

Anyone who cares not about his ancestors deserves to be forgotten by his descendants or has no future as an ancestor. The world is full of anecdotes about forebears. We often ascribe to ancestors abilities they did not have, honors they did not win, fortitudes they had, or inabilities to get along where they lived. Most were dissatisfied because they didn't fit in their European environments, so they left and went elsewhere. Yet they must be admired because they did something about their dissatisfaction. As we look back and try to imagine the conditions under which they lived, we go to countries from which they came, not realizing that the people who stayed behind have also evolved, as have we, into a people vastly different from those our ancestors left when they came to America.

In 1977, my wife, Florence, and I spent a whole summer back East among the brambles of lost cemeteries, digging up old downed headstones from the 17th and 18th centuries, bedeviling town clerks and village librarians, and rummaging through the musty records of local historical societies in pursuit of ancestors long moldered to dust. We ranged through Massachusetts, Connecticut and New York State photographing tombstones of my wife's Puritan and Pilgrim ancestors, checking dates of birth, dates of death and correct names found in the marvelous genealogical library of the Mormons in Salt Lake City. In some counties, cemeteries have been catalogued, tombstones read and publications produced which greatly aided our search. In Connecticut alone, Florence drove 1600 miles in a month, crisscrossing that tiny state to chase down cemeteries and local records. Never had anyone else been able to travel that many miles in that small an area in such a short time. Her tenacity was amazing, and she became a sort of local celebrity as she roared around Connecticut in her little red Honda while I stayed in our motorhome writing on projects of my own. Town and county clerks, directors of local museums, clerks in historical societies, and ordinary citizens yielded to her requests for information, even going so far as to show us cemeteries long hidden in dense woods, forgotten on hills, and buried in cultivated fields and on grassy slopes. With maps we surveyed locations where gravestones should have been but are now long gone, perhaps vandalized or used by local farmers as foundations for barns, which are also gone. Most of the truly old graves had headstones no longer legible, but they had been read scores of years before so that records were preserved and available.

Ancestral church of the Mather family
Winwick, England

As a dedicated genealogical researcher, Florence Mather Wengerd has no peer. Fortified with published data and an old detailed map book showing villages, towns and cities of England, we visited 117 ancient Anglican churches scattered over the whole of that historic country. We searched every nook and cranny, driving 6500 miles in just two months during 1990. We stayed at pubs—those marvelous oases of food, drink and lodging so endemic to England, even in the ancient days of her ancestors in the 14th through the 17th centuries. These churches were marvels of ancient stone construction, the stone usually having been quarried locally. One could make a geologic map of England just by studying the rock used to construct and reconstruct churches built since 1200 AD. These churches all had grand old bells cast in Great Britain. Some bells had been removed and placed in town squares, and other early bells had been purloined by later churches. The lead roofs had often been stolen by thieves for resale. Most, however, were in various states of preservation by local church societies who recognized their historical value. One very old church was a virtual ruin, a heart-breaking situation with no committee, society or town to save it. Most were still in use as local churches. Many were built on sites of Celtic and Roman churches in existence as early as 800 and 900 AD before the Normans, under William the Conqueror, came in 1066 AD. All were the oldest Anglican churches in the religious renaissance, founded when Henry VIII, excommunicated by the Pope because of his penchant for changing wives, simply started his own church. Because they were the oldest, and in many cases the only church in many a community, many could be located by finding a street named Church Lane or Church Street. These were the churches where Florence's ancestors worshipped, were married, were consecrated, died, and were buried. Some 15 of these ancient edifices had plaques, stained glass windows, stone tablets, or some other types of physical evidence illustrating her direct ancestors' participation in the histories of these churches. No gravestones made of slate, limestone, gritstone, or sandstone could sustain names because of rigorous weathering through the centuries, though written records proved that her ancestors were intimately connected with every one of the 117 churches we visited. Florence's records will eventually be published. They involve hundreds of families with illustrious backgrounds of worthy, nonconforming yeomen whose descendants peopled much of early America.

The old cemeteries of the eastern United States have almost countless gravestones with interesting phrases, poems, and epitaphs to the people buried there. The old ones show carvings in bas relief of crossbones, skulls, the death mask, and doves. On one was an especially poignant epitaph to two friends buried side by side, friends who never took time to really visit with each other. It goes thus: "In death united, in life too busy." Others include "Heed ye passerby, as he who lies here, so will you too be," and, on the gravestone in Ohio of the well-known professor of music, Sam Gelfer, and his wife, Inez, reads "Music is love in search of a word." Over a beautiful grave, a stone on the atoll of Majuro in the Marshall Islands reads "Here he lies, where he ought to be." Obviously an epitaph to a wayward husband not missed!

McCauley, in his book *History of England*, makes this pithy comment: "A people which takes no pride in the noble achievements of remote ancestors will never achieve anything worthy to be remembered with pride by remote descendants." Much the same idea by Edmund Burke seems to prove the concept that "People will not look forward to their posterity who will not look back to their ancestry."

The pursuit of ancestors is intensely habit-forming and never ending. Beside that, it can be costly. However, such exploration leads to valuable historical discovery, and it is surely satisfying.

The Aging Syndrome

It is not an idle exercise to speculate on how long one may live. Doctors will tell you one measure—highly suspect—is to average the ages of your grandparents. Mine average 78 years. Another is to fill out one of these computer-generated questionnaires based on statistical analyses of health items involving thousands of people who have died of various causes, plus a survey of personal habits. Mine comes out to 79 years. Another is to do a statistical analysis of the number of people today who are at a certain age and then project what a terminal age might be, based on an average age of death as a moving average. Mine comes to 86 years. Now that's better but equally fallacious, as are all such computations!

There is not a man or woman alive who will not die. A Catholic padre said it best: "My son, never fear death. Since man has lived, who has not died?" Death, though inevitable and the ultimate tragedy, comes as a welcome transition to conditions unknown if one has lived a good life. It has always amazed me how many religious people want to go to heaven and, yet, do not want to die. Heaven is not a place but a condition, and it's a difficult conversation topic, especially if one is convinced that heaven is here and now—make the best of it.

Few thinking persons have not thought of aging. Many have tried unsuccessfully to stop the process. Just today I came up with the thought that all ship voyages, of any length, do not count against the span of one's life. Stated another way, if one travelled at sea a third of one's life, one would live that much longer—a pregnant thought which all would agree is a good idea. As we go through life trailing our nerves, shattered by noise, pollution, worry, and the perfidy of humans, it's good to remember that this too shall pass and, while passing, we should seek simplicity, as Thoreau counseled.

As we pass this way, trying to make the world a better place than we found it, there are worries enough without doting on aging. Mike Schmidt said, "If it isn't my wife or my life, then why the heck should I worry?" A friend, Karl Albert, now long gone and far too soon, thought, "No matter what happens they can't eat you up!" All of life is a love affair with the harsh mistress of aging, and there doesn't seem to be much use in getting so all-fired serious in this life. You're not going to get out of it alive anyway. When death does come, 99 percent of the time it's not due to aging but to disease, accident or the will to die. Of course aging makes everyone more susceptible to disease, and accidents are generally due to one's own careless behavior. However, a strong will to die, whether by hex, voodoo, or a human's natural characteristic to be pessimistic, is truly self-inflicted. I am convinced that aging can be defeated until the last day of one's life if one simply remains upright, laughing, mobile, more than half smart, and completely optimistic that it won't rain or snow. Fred Allen, in answer to the question whether he believed in the hereafter, said, "You only live once, but if you live it right, once is enough." So much for a desire of reincarnation.

There are people who never considered that they were getting old. Professor Kirtley Mather talked of his advancing maturity (he lived to be 90); his sister Juliette considered her receding youth and died at 87. Lewis Wright said, "Old, what the heck do you mean old? I've just lived a lot of years!" Tom Collins maintained that there are three ages for man: "Youth, maturity, and you haven't changed a bit." Yet I have friends who continually talk about being old, and their ages range from sixty to seventy. My son thought I was old when I was forty. There is, however, a Spanish proverb that urges caution: "*Debemos respetar las canas, especialmente nuestras*." (We ought to respect gray hairs, especially our own.) A recognition of limitations is a prime ingredient in aging gracefully. One of the sure characteristics of age is a shuffling gait, so step along if it kills you! There's nothing so sad as an old roué or an aging woman who dresses like a teenager. You are what you think, but be careful that your thoughts match your age when it comes to strenuous physical activity. One of the major benefits of a lengthening life is that you don't have to prove anything to anybody. You can jolly well do as you please as long as you hurt no one. You can spend all the money you have, if you are so inclined, and you don't give a darn about tomorrow.

A very good friend of mine, Dr. Martin Fleck, raconteur extraordinaire and retired professor of biology at the University of New Mexico, averred that his health was good for three main reasons. He said, "I had good parents, I don't waste energy resisting temptation, and I don't eat natural foods. At my age of 86, I need all the preservatives I can get." He also thought he could understand why he wasn't over the hill: "It is because I haven't gotten to the top yet."

La Rochefoucauld touched a nerve with "Old people love to give good advice; it compensates them for their inability to set a bad example." Nestor Pestelos of Fiji put it this way: "We grow old when regrets take the place of dreams." Expressions of love and feeling must be timely. Witness Alton Collins: "You can never do a kindness too soon, for you never know how soon it will be too late." From all of this, one must distill for himself the true essence of successful aging.

Colorado River trip
Cataract Canyon
1971

The Relentless Tide

A first brush with death is a shock one never forgets. When I was but a lad of eight, a maiden aunt of mine died. She was the kindliest of souls with a mind not quite right. She was taken out of the South Bunker Hill country school at the age of ten, unable to progress past the third grade. Her mother, Catherine Miller, and father, Alexander E. Miller, were second cousins—and were also my mother's parents. Her name was Amanda, and we called her Mandy. She was a favorite of all her nieces and nephews of whom there were many because she had three brothers and six sisters. She was born in 1884, the third child in a large family of German Swiss Mennonites.

When Mandy died I suffered grievous shock. We held the funeral in the Bunker Hill Dunkard Church used by local Mennonite families. The small wooden church was filled to overflowing, people covering the lawn outside and horses with buggies tied up everywhere. Amish and Mennonites from all around came to her funeral. Mandy, mentally damaged before birth, was considered by all who knew her to be a saint. Through her short life of 39 years, her mother trained her to cook and do housework, and my grandfather taught her how to work in the fields just as he taught all his daughters and sons.

We all cared very deeply for this lovely person. She was small, not at all pretty, could speak only Pennsylvania Dutch, and was slow moving, but would play games with us by the hour. The night she lay dying she saw an angel in an upper corner of her small bedroom, and, as we looked at her in her vision, we could see angels in her eyes. Baptized in the Mennonite church when she was 12 years of age, she read the Bible in German, prayed with the same vigor she played with us, and was the epitome of religious fervor. The undertaker came to embalm her the same evening she died—no doctor was available—and my grandmother sat by her still body through the night. All of the family was called, and the funeral was held on a Sunday, just two days after Mandy had died. I had tried to see her while she lay dying, but my grandfather would not allow it until she was dressed in her finest dress, which had been made by my grandmother.

Seven years after Mandy's death, Grandpa suffered a stroke. That Sunday both my grandma and my grandpa had escaped the gathered family with Old Bert hitched to their buggy to visit some friends in the Walnut Creek Valley some seven miles to the south. Far too early, we saw Grandma driving the buggy at a wild pace up the road back to the farm. Alarmed, we ran out to find Grandpa slumped over, unconscious. The next day a second stroke killed Grandpa Aleck, as we had so long called him with affection. Now an even larger funeral took place with burial in the lovely South Bunker Hill Cemetery. Only six years later, Grandma was felled by two strokes, and the large, saddened family gathered again, this time to say good-bye to Catherine, the saint of the large Miller family. I was to experience the death of my older brother, my parents and many friends and relatives through the almost 70 years after we lost Aunt Mandy.

There is a relentless tide in families as death marches onto the scene to sadden lives. Preachers make a big to-do about Heaven, a life hereafter, the re-gathering of families, on and on, thinking to make everything right in a future after death, a future about which no one knows. The tide rolls on—maybe we should call it a tsunami in the fitful destruction known as death. But if there was no death, there would be no philosophical outlooks about how precious life really is. Death solves many seemingly insoluble problems and leads to some of our greatest poetry, our most poignant stories, our most thrilling operas, and our deepest emotions.

To experience the death of a close relative as a very young boy, disheartening as it is, creates in one wonder at how grand life really is. To contemplate and experience such loss is a reminder that heaven is most likely here and now.

Our last trip—
November 7, 1994

Death Anytime

The prime fascination with death is that it happens to us all sooner or later. Concepts of death permeate poems and prose in all lands as well as religions, spoken and taught. It is feared because no one has returned to tell us about it. It has been likened to going through a door between rooms, leading to strong religious inferences that another room exists, that somehow we will all come back through doors at the same time after Armageddon. Some view it as a punishment for sins, others as a reward; some view death as our just desserts for being so nasty to each other. Time wounds all heels!

There appear to be more quotations about death than any other subject except sex and love—both must be more enjoyable because they occur here for all to experience. Elsewhere I've recorded the question of Father Hoa: "My son, never fear. Since man has lived, who has not died?" If one could choose a place to die, one would chose a place where one belonged, feeling comfortable in adversity and dying in peace, even though one's work is undone. A sailor's prayer includes this: "Lord, let me die in a hammock"—a comfortable thought!

At the age of 48, my father thought he was dying of tularemia; as he turned his face to a wall, alone, he beseeched God to let him live long enough to see his youngest child, his only daughter, live to womanhood. The prayer was answered, one supposed, for he lived until he was almost 80 years of age. Obviously, such experiences may make people, previously lukewarm about God and Heaven, quite religious.

Wise men, unsure like the rest of us about the existence of an afterlife, believe death to be the ultimate tragedy. It is not unusual to think that "If I were to die tonight, will anybody be able to figure out the mess I've left behind?" The spiritual side of any thinking human can take solace and be cheerful of heart and be patient. God hasn't finished with us until we die—and maybe not even then. As one visits Arlington Cemetery with its thousands of white crosses designating the graves of service men killed in their youth, one can only be sad. But one thought comes to mind: "They will never grow old!" Inasmuch as it can happen anytime, rejoice each day that you are alive. The days may be numbered, but by whom or what no one knows.